I0502860

Standard Review Plan
For Releasing Part
Of a Reactor Facility or Site
For Unrestricted Use
Before Approval
Of the License Termination Plan

Draft Report for Comment

Manuscript Completed: February 2006
Date Published: October 2006

Prepared by
J-C. Dehmel

Division of Fuel, Engineering and Radiological Research
Office of Nuclear Regulatory Research
U.S. Nuclear Regulatory Commission
Washington, DC 20555-0001

COMMENTS ON DRAFT REPORT

Any interested party may submit comments on this report for consideration by the staff of the U.S. Nuclear Regulatory Commission (NRC). Comments may be accompanied by additional relevant information or supporting data. Please specify **Draft NUREG-1836** in your comments, and send them — by the end of the 60-day comment period specified in the *Federal Register* notice announcing availability of this draft — to the following address:

Chief, Rules Review and Directives Branch
Mail Stop: T6-D59
U.S. Nuclear Regulatory Commission
Washington, DC 20555-0001

You may also submit comments electronically using the NRC's public Web site:

http://www.nrc.gov/public-involve/doc-comment/form.html

For any questions about the material in this report, please contact:

Jean-Claude Dehmel
Mail Stop: T9-C34
U.S. Nuclear Regulatory Commission
Washington, DC 20555-0001
Phone: (301) 415-6619 or (800) 368-5642
Email: JXD3@nrc.gov

ABSTRACT

This standard review plan (SRP) provides guidance to the staff of the U.S. Nuclear Regulatory Commission (NRC), Office of Nuclear Reactor Regulation and Office of Nuclear Material Safety and Safeguards, on how to evaluate requests to release part of a power reactor facility or site for unrestricted use prior to NRC approval of the facility's license termination plan (LTP). This type of release is referred as a "partial site release" (PSR). The associated regulatory requirements are contained in Title 10, Section 50.83, "Release of Part of a Power Reactor Facility or Site for Unrestricted Use," of the *Code of Federal Regulations* (10 CFR 50.83). This SRP presents evaluation and acceptance criteria for the review of PSR applications submitted by licensees. In so doing, this SRP identifies technical areas for review, information to be submitted by licensees, regulatory requirements and guidance, and evaluation criteria for use in assessing compliance with the regulations. Although the main focus of this guidance is directed to the NRC staff's review process, this SRP presents information that licensees and applicants may find useful. Another important purpose of the SRP is to make information available so that the public has a better understanding of the staff's review process and criteria for rejecting or accepting such requests.

Paperwork Reduction Act Statement

Public Protection Notification

Notice

FOREWORD

Under current regulations, a portion of a power reactor site may be released from regulatory control before the U.S. Nuclear Regulatory Commission (NRC) terminates the license. This type of release is defined as a "partial site release" (PSR). NRC regulations, under Title 10, Section 50.83, "Release of Part of a Power Reactor Facility or Site for Unrestricted Use," of the *Code of Federal Regulations* (10 CFR 50.83), present requirements and a process that a licensee would use to request approval for a PSR from the NRC. These requirements are separate from those of 10 CFR 50.82, "Termination of License," which address decommissioning and license termination. As a Standard Review Plan (SRP), this document provides guidance for NRC staff to use in reviewing applications for PSR. Specifically, this SRP identifies technical areas for review, information to be submitted by licensees, regulatory requirements and guidance, evaluation criteria for use in assessing compliance with regulations, and references. Although the main focus of this guidance is directed to the NRC staff's review process, this SRP presents information that licensees and applicants may find useful. Another important purpose of the SRP is to make information available so that the public has a better understanding of the staff's review process and criteria for rejecting or accepting such requests.

At this time, the NRC is issuing this draft SRP to solicit comments from the public and other stakeholders. The NRC encourages participation in the development of this document and implementation of the associated rule. The use of this SRP is not authorized until it has been finalized by the NRC.

Brian W. Sheron, Director
Office of Nuclear Regulatory Research
U.S. Nuclear Regulatory Commission

CONTENTS

EXECUTIVE SUMMARY

Under current regulations, a portion of a power reactor site may be released from regulatory control before the U.S. Nuclear Regulatory Commission (NRC) terminates the license and before the licensee submits a license termination plan (LTP). This type of release, defined as a "partial site release" (PSR), may involve radiologically impacted or non-impacted areas. The regulations set forth in Title 10, Section 50.83, "Release of Part of a Power Reactor Facility or Site for Unrestricted Use," of the *Code of Federal Regulations* (10 CFR 50.83), present requirements that identify the criteria and regulatory framework that a licensee would use to seek NRC approval of a PSR request. These requirements apply only to PSRs involving unrestricted use of the area following release. As such, 10 CFR 50.83 focuses only on PSRs for operating reactors and those being decommissioned before approval of the LTP. These requirements are separate from those of 10 CFR 50.82, "Termination of License," which address decommissioning and license termination. The License Termination Rule (10 CFR Part 20, Subpart E) provides requirements and criteria for acceptable dose to average members of the critical group resulting from residual radioactivity remaining in structures, materials, soil, groundwater, and other media at a reactor site after the reactor license has been terminated. Subpart E contains the radiological criteria for releasing an entire reactor site following decommissioning. However, Subpart E does not present criteria regarding impacts associated with the decommissioning work itself. The regulations on PSR require licensees to submit information necessary to demonstrate the following:

- The release of a portion of a radiologically impacted area satisfies the radiological criteria for unrestricted use in 10 CFR 20.1402, as they relate to the acceptable dose [0.25 mSv/yr (25 mrem/yr)] to the average member of the critical group, including the provisions for doses that are as low as is reasonably achievable (ALARA).

- For impacted or non-impacted areas, the release of the property has no adverse effects on reactor operations, siting, and safety.

- Doses to individual members of the public do not exceed the limits of Subpart D to 10 CFR Part 20 for either impacted or non-impacted areas.

- The licensee shall continue to comply with all other applicable regulatory requirements that may be affected by the release of property and changes to the site boundary. These requirements address site operations, technical specifications, radiological effluents, emergency planning, site security, environmental monitoring, and siting criteria. These requirements are addressed in 10 CFR Parts 20, 50, 72, and 100.

- Records of property line changes and the radiological conditions of partial site releases are maintained to ensure that the dose from residual material associated with these releases can be accounted for at the time of any subsequent partial releases and at the time of license termination.

This standard review plan (SRP) provides guidance to NRC staff for use in reviewing and evaluating applications for PSRs. Specifically, this SRP identifies technical areas for review, information to be submitted by licensees, regulatory requirements and guidance, evaluation criteria for use in assessing compliance with regulations, and references. The approval process by which the property would be released depends on the radiological classification of the area to be released, defined as an "impacted area" or "non-impacted area" in 10 CFR 50.2.

Further details are provided in NUREG-1575, "Multi-Agency Radiation Survey and Site Investigation Manual" (MARSSIM). For areas classified as "non-impacted" and, therefore, having no reasonable potential for residual radioactivity and for which release would not adversely affect reactor operation and safety, the staff would approve the release by letter. For areas classified as "impacted" and, therefore, having some reasonable potential for residual radioactivity, the regulations specify that the licensee must submit a license amendment request for NRC review and approval. That amendment request would include the licensee's demonstrated compliance with the radiological criteria for unrestricted use specified in 10 CFR 20.1402. Additional regulatory guidance for evaluating requests for PSRs and license amendments is contained in NUREG-1757, "Consolidated NMSS Decommissioning Guidance," dated September 2003.

ACKNOWLEDGMENTS

This standard review plan (SRP) was developed through the collaboration of staff from several offices of the U.S. Nuclear Regulatory Commission (NRC). The author wishes to recognize the efforts of Mr. W.M. Ripley from the NRC's Office of Nuclear Reactor Regulation (NRR), who developed the first working draft of this document, and Mr. H.S. Tovmassian (NRR), who assisted in assembling material supporting the preparation of the first draft. In addition, the following technical reviewers provided valuable comments and recommendations to support the preparation of this SRP.

R. Abu-Eid [Office of Nuclear Material Safety and Safeguards (NMSS)]
C. Craig (NMSS)
S. Bush-Goddard [Office of Nuclear Regulatory Research (RES)]
J. Hickman (NMSS)
S. Klementowicz (NRR)
C. McKenney (NMSS)
G. Mizuno [Office of the General Counsel (OGC)]
D. Orlando (NMSS)
D. Schmidt (NMSS)
D. Sollenberger [Office of State and Tribal Programs (STP)]

ABBREVIATIONS

alpha (α)	a MARSSIM term for a specified probability of a Type I error
ADAMS	Agencywide Documents Access and Management System
ALARA	as low as is reasonably achievable
beta (β)	a MARSSIM term for a specified probability of a Type II error
Bq	becquerel, a unit of radioactivity
BTP	branch technical position
CERCLA	Comprehensive Environmental Response, Compensation, and Liability Act
CFR	*Code of Federal Regulations*
cm^2	square centimeter
DandD	a screening radiological assessment computer code for site grounds (superficial soils) and buildings
DCGL	derived concentration guideline level
$DCGL_W$	DCGL for the nonparametric statistical test
D&D	decontamination and decommissioning
DIRS	Division of Inspection and Regional Support (NRR)
DNMS	Division of Nuclear Materials Safety (NRC regional offices)
DORL	Division of Operating Reactor Licensing (NRR)
dpm	disintegration per minute
DPR	Division of Policy and Rulemaking (NRR)
DQI	data quality indicator
DQO	data quality objective
DRA	Division of Risk Assessment (NRR)
DRP	Division of Reactor Projects (NRC regional offices)
DRS	Division of Reactor Safety (NRC regional offices)
DWMEP	Division of Waste Management and Environmental Protection (NMSS)
EA	environmental assessment
EAB	exclusion area boundary
EIS	environmental impact statement
EPA	U.S. Environmental Protection Agency
ER	environmental review
FSAR	final safety analysis report
FSS	final status survey
FSSR	final status survey report
g	gram
HSA	historical site assessment
ISFSI	independent spent fuel storage installation

LPZ	low population zone
LTP	license termination plan
MARLAP	Multi-Agency Radiological Laboratory Analytical Protocols (NUREG-1576)
MARSSIM	Multi-Agency Radiation Survey and Site Investigation Manual (NUREG-1575)
MDA	minimum detectable activity
MDC	minimum detectable concentration
MOU	memorandum of understanding
mSv	millisievert, a unit of dose
mrem	millirem, a unit of dose
NMSS	Nuclear Material Safety and Safeguards, Office of (NRC)
NRC	U.S. Nuclear Regulatory Commission
NRR	Nuclear Reactor Regulation, Office of (NRC)
OCA	Office of Congressional Affairs (NRC)
ODCM	offsite dose calculation manual
OGC	Office of the General Counsel (NRC)
OPA	Office of Public Affairs (NRC)
PARCC	precision, accuracy (bias), representativeness, comparability, and completeness
pCi	picocurie, a unit of radioactivity
PM	project manager
PSR	partial site release
QA	quality assurance
QC	quality control
RAI	request for additional information
REMP	radiological environmental monitoring program
RES	Nuclear Regulatory Research, Office of (NRC)
RESRAD	a radiological assessment computer code for site grounds
RESRAD-BUILD	a radiological assessment computer code for facility buildings
RIS	regulatory issue summary
SER	safety evaluation report
SFPO	Spent Fuel Project Office (NMSS)
SRP	standard review plan
STP	State and Tribal Programs, Office of (NRC)
USAR	updated safety analysis report
WRS	Wilcoxon Rank Sum

1. INTRODUCTION

On April 22, 2003, the U.S. Nuclear Regulatory Commission (NRC) amended its regulations[1] to standardize the process for allowing a nuclear power reactor licensee to release part of its facility or site for unrestricted use before the NRC approves the license termination plan (LTP). This type of release is referred as a "partial site release" (PSR). The amendment adds a new section [Title 10, Section 50.83, "Release of Part of a Power Reactor Facility or Site for Unrestricted Use," of the *Code of Federal Regulations* (10 CFR 50.83)][2] identifying the criteria and regulatory framework that a licensee would use to seek NRC approval for a PSR request. The regulations impose parameters for unrestricted use of the radiological criteria for license termination specified in Subpart E of 10 CFR Part 20, "Standards for Protection Against Radiation," for PSRs and exclude PSRs for restricted uses. The regulations also provide additional assurance that residual radioactivity at the site would meet the radiological criteria for license termination, even if parts of the site were released before NRC approval of the LTP.

The regulations set forth in 10 CFR 50.83 do not apply to byproduct material, source material, or non-power reactor licensees. Partial site releases considered after LTP approval for restricted or unrestricted use would be addressed in the LTP, if it contained such provisions and a process for staged releases, or through an amendment to the LTP. Additional guidance for obtaining NRC approval for PSRs after LTP approval is provided in NUREG-1700, NUREG-1757, and NUREG-1575 (NRC 2003a, 2003b, and 2000).

The regulations require licensees to submit specific information to demonstrate the following:

- For impacted areas, the licensee must comply with the radiological criteria for unrestricted use, as specified in 10 CFR Part 20, Subpart E (10 CFR 20.1402) [0.25 mSv/yr (25 mrem/yr) and as low as is reasonably achievable (ALARA) provisions].

- For impacted or non-impacted areas, the release of the property must have no adverse effects on reactor operations, siting, and safety.

- Doses to individual members of the public must not exceed the limits of 10 CFR Part 20, Subpart D, for impacted or non-impacted areas.

- The licensee must continue to comply with all other applicable regulatory requirements of 10 CFR Parts 20, 50, 72, and 100, as specified in the license and final or updated safety analysis report (FSAR or USAR). These requirements address site operations, technical specifications, radiological effluents, emergency planning, site security, environmental monitoring, and siting criteria that may be affected by the release of property and changes to site boundaries. These requirements apply to both impacted and non-impacted areas.

---

[1]Final Rule, "Releasing Part of Power Reactor Site or Facility for Unrestricted Use Before the NRC Approves the License Termination Plan," *Federal Register*, Vol. 68, No. 77, pp.19711–19728, April 22, 2003.

[2]Full citations of the *Code of Federal Regulations* are given in the references at the end of this section.

- The licensee must maintain records, in accordance with 10 CFR 50.75(g), documenting all related changes in site property boundaries and radiological conditions of PSRs to ensure that doses from residual radioactivity associated with all releases can be accounted for at any subsequent partial site releases and at the time of license termination for the entire site. These requirements apply to both impacted and non-impacted areas.

The review process by which a part of the property would be released depends on the potential for residual radioactivity remaining in the area, defined in the rule as "non-impacted areas" or "impacted areas." The following summarizes the regulatory framework that a licensee or applicant shall use to submit a PSR request, and steps the NRC staff shall use to review, evaluate, and process such a request:

- For non-impacted areas:
  - The applicant shall comply with 10 CFR 50.83(a)(1) and (a)(2) and 10 CFR 50.83(b)(1) – (b)(5).
  - The NRC review, evaluation, and processing steps are defined in 10 CFR 50.83(c) and 10 CFR 50.83(f).
  - The applicant shall comply with applicable parts of the standard review plan (SRP) (Sections II.1 – II.4, II.7, and II.8).
- For impacted areas:
  - The applicant shall comply with 10 CFR 50.83(a)(1) – (a)(3) and 10 CFR 50.83(d)(1) – (d)(3).
  - The NRC review, evaluation, and processing steps are defined in 10 CFR 50.83(e) and 10 CFR 50.83(f).
  - The applicant shall comply with applicable parts of the SRP (Sections II.1 – II.3 and II.5 – II.8).

The definitions of "impacted areas" and "non-impacted areas" are included in the SRP and in 10 CFR 50.2, while additional details are provided in NUREG-1575, "Multi-Agency Radiation Survey and Site Investigation Manual (MARSSIM)." For areas classified as "non-impacted" and having no reasonable potential for residual radioactivity, the staff may approve the release of the property by letter, provided that the release of the property has no adverse effects on reactor operations, siting, and safety. For areas classified as "impacted" and having some reasonable potential impact attributable to residual radioactivity, the regulations require licensees to submit a request for a license amendment. The amendment request must demonstrate the licensee's compliance with the radiological criteria for unrestricted use [0.25 mSv/yr (25 mrem/yr) to the average member of the critical group and the ALARA provision, as specified in 10 CFR 20.1402]. If an impacted area was remediated to meet the dose criteria of 10 CFR 20.1402, the amendment request should include this information in the historical site assessment (HSA). This information should describe the original radiological conditions of the area before remediation, provide the technical basis for residual radioactivity criteria corresponding to the dose limit of 10 CFR 20.1402, and provide radiological survey results demonstrating compliance with residual radioactivity criteria. License amendment requests for impacted areas shall include a supplement to the environmental report, under 10 CFR 51.53, describing any new information or significant environmental change associated with the proposed PSR.

Power reactor licenses incorporate conditions and technical specifications with detailed descriptions of the site boundary, e.g., a site map identifying areas controlled by the licensee, restricted industrial areas, and radiologically controlled areas. If the licensed site boundary changes as a result of a PSR, licensees shall request a license amendment regardless of the radiological classification of the area and new configuration of the site boundary.

The regulations include provisions for public participation. The staff shall notice receipt of a licensee's proposal for a partial site release in the Federal Register, regardless of the potential for residual radioactivity, and make the licensee's application available for public review and comment. Also, the staff shall hold a public meeting in the vicinity of the site to discuss the licensee's request or license amendment application, as applicable, and obtain comments before acting on the application.

1.1   Review Phases and Responsibilities

The NRC's review and evaluation process for a proposed PSR application consists of six major steps:

(1)   Acceptance Review: Determine whether the information submitted by the licensee is complete and sufficiently detailed to support a technical and regulatory review.

(2)   Public Participation: Make the licensee's request available to the public, schedule a public meeting in the vicinity of the facility, and review and resolve public comments.

(3)   Detailed Technical Review: Initiate the review of the application package, including requests for additional information (RAIs), as needed, to support the staff's evaluation and determination of acceptability.

(4)   Site Inspections and Confirmatory Surveys: Inspect the area that the licensee proposed for PSR, review records supporting the application, conduct a radiological survey (as needed) of the area to be released, and assess potential mechanisms for transporting radioactive contaminants from the licensed portions of the site to the area designated for release.

(5)   Evaluation Findings: Determine the acceptability of the application and confirm compliance with regulatory requirements.

(6)   Approval Process: Prepare and issue, as required, an approval letter, a license amendment, an environmental assessment (EA) or environmental impact statement (EIS), and a safety evaluation report (SER).

The cognizant NRC project manager (PM) for the licensed facility has the responsibility for initiating, coordinating, and scheduling all reviews, public meetings, and site inspections, as well as documenting the staff's findings and recommendations. The PM may be a project manager in the Office of Nuclear Reactor Regulation (NRR), or the Office of Nuclear Material Safety and Safeguards (NMSS) if the facility is in decommissioning and has been transferred to NMSS.

In some instances, the PM may require technical assistance from other NRC headquarters and regional offices, divisions, and branches. In general, the following NRC offices and divisions[3] may be involved in the review process:

- Office of Nuclear Reactor Regulation (NRR)

    - Division of Inspection and Regional Support (NRR/DIRS)
    - Division of Operating Reactor Licensing (NRR/DORL)
    - Division of Risk Assessment (NRR/DRA)
    - Division of Policy and Rulemaking (NRR/DPR)

- Office of Nuclear Material Safety and Safeguards (NMSS)
    - Division of Waste Management and Environmental Protection (NMSS/DWMEP)
    - Spent Fuel Project Office (NMSS/SFPO)

Office of Public Affairs (OPA)

- Office of Congressional Affairs (OCA)

- Office of the General Counsel (OGC)

- Regional Offices
    - Division of Reactor Projects (DRP)
    - Division of Reactor Safety (DRS)
    - Division of Nuclear Materials Safety (DNMS)

The specifics of each application will dictate which functional branches and technical expertise are needed to conduct the appropriate review and evaluation. The PM is responsible for coordinating the support of other Offices and Divisions in conducting specific technical reviews, as needed. The PM is also responsible for integrating the results of these reviews into the overall evaluation and determination of findings.

---

[3]See the NRC's public Web site (http://www.nrc.gov) for a description of the agency's organizational structure and functions.

1.1.1 Acceptance Review

Before performing the technical review, the PM and assigned technical reviewers shall determine whether the application involves an impacted or non-impacted area, and define the scope of the acceptance review. This review is performed to determine whether the information submitted by the licensee meets the requirements of 10 CFR 50.83(a) and (d) for impacted areas, and 10 CFR 50.83(a) and (b) for non-impacted areas. The acceptance review shall follow the requirements of 10 CFR 50.83(c) and (f) for a non-impacted area, and 10 CFR 50.83(e) and (f) for an impacted area, which requires the issuance of a license amendment. This review should be completed within 30 days. Upon completion of the acceptance review, the PM shall forward an acknowledgment letter to the licensee and inform the public of the licensee's request by publishing a notice in the *Federal Register* and local news media. (See the discussion of public participation and coordination in the next section.)

1.1.2 Public Participation and Coordination

Under 10 CFR 50.83(f), the NRC is required to publish notices of receipt of all requests for PSR approval or related license amendments, and make those requests available for public comment. In addition, the procedures of 10 CFR 50.91 shall be considered whenever public or State participation are mandated. The purpose of public meetings is to address substantive issues that are directly associated with the NRC's regulatory responsibilities in light of the proposed request. If warranted, the PM shall inform other regulatory agencies or local government entities that have an interest in the site or a regulatory role on the proposed action. Public meetings will be held as Category 3 meetings.

If the application is found to be complete, the PM shall forward an acknowledgment letter to the licensee and inform the public of the licensee's request by publishing notices in the *Federal Register*, on the NRC's public Web site, and in local news media outlets that are readily accessible to the public in the vicinity of the site. Also, the PM is responsible for scheduling a public meeting in the vicinity of the site to discuss the licensee's request and solicit stakeholder comments. The NRC shall announce in a *Federal Register* notice, on the NRC's public Web site, and in local news media outlets the purpose of the public meeting, and its scheduled date, time, and location. The meeting will be announced no fewer than 10 days before the meeting date. The licensee's request shall be made available to the public by placing the application in the NRC's Agencywide Documents Access and Management System (ADAMS) under the appropriate docket number. The ADAMS accession number shall be included in the *Federal Register* notice.

In the event that a PSR-related license amendment is challenged, an opportunity for a hearing shall be provided under the provision of 10 CFR 2.206. This provision allows members of the public to raise public health and safety concerns and petition the NRC to take specific actions to resolve concerns identified in the petition.

Stakeholder participation shall be addressed in accordance with the related NRC guidance and policy statement. NRC guidance for establishing and conducting meetings is contained in NUREG-1757, Vol. 1, Section 4.8. The NRC policy statement, "Enhancing Public Participation in NRC Meetings: Policy Statement," can be found in the *Federal Register* (Vol. 67, No. 102, p.36920, May 28, 2002). Comments from stakeholders shall be reviewed and evaluated by the NRC before acting on the request for a PSR.

1.1.3 Detailed Technical Review

Reviewers shall perform a technical review to determine whether the licensee's request complies with the requirements of 10 CFR 50.83, whether the remaining areas of the site or site operations would be affected as a result of the release or lead to changes in the site boundary, and whether the facility or site meets the radiological release criteria for unrestricted use, as set forth in 10 CFR Part 20, Subpart E, "Radiological Criteria for License Termination." The review should verify that the radiological characterization of the site is properly described, and that the licensee's conclusions as to the radiological designation and survey classification of the area are supported by the HSA, site characterization data, and process contained in the relevant sections of NUREG-1555, NUREG-1575, NUREG-1576, and NUREG-1757 (NRC 1999, 2000, 2003b, 2004). The technical review shall address the following considerations:

- general information and description of the area proposed for PSR

- characterization of the area in determining whether the area is impacted or non-impacted by current and prior site operations

- impacts on the licensee's programs, including site operations, technical specifications, radiological effluent releases and offsite doses, environmental monitoring, emergency plan, security plan, and siting criteria

- calculation methods and results demonstrating that (1) the measured concentrations of residual radioactivity yield doses to average members of the critical group in compliance with release criteria, or (2) dose criteria were used to calculate derived concentration guideline levels (DCGLs), which are used as limits in evaluating final status survey (FSS) results

- results of the FSS performed by the licensee

- manner in which the PSR release criteria will be considered during the development of the final site criteria at the time of license termination for the entire site

- basis for radiological criteria for unrestricted release

- HSA and operational records forming the basis of the request for the PSR

If groundwater or soil is known or suspected to be contaminated with plant-derived radioactivity, the PM shall consider the provisions of the memorandum of understanding (MOU) between the U.S. Environmental Protection Agency (EPA) and the NRC to identify their respective roles for decommissioning NRC-licensed sites (*Federal Register*, Vol. 67, No. 206, p.65375, Oct. 24, 2002). The MOU states that the EPA will defer its authority under the Comprehensive Environmental Response, Compensation, and Liability Act (CERCLA) for the majority of facilities decommissioned under NRC authority. The MOU includes provisions for joint consultations for sites when, at the time of license termination, (1) groundwater contamination exceeds EPA maximum concentration levels, (2) the NRC contemplates restricted release or alternative criteria for the site, or (3) residual radioactive soil concentrations exceed levels defined in the MOU.

The evaluation of PSR requests for facilities and sites that are known or suspected to have contaminated soil or groundwater shall address whether any of the following criteria exist:

- The portion of the site designated for release is eligible for a PSR in its current or expected radiological condition, given surface and subsurface transport mechanisms.

- Remediation might be a feasible option in mitigating the presence of contaminated soil and groundwater.

- The PSR action might be deferred to the time of license termination and implemented as a staged release under an approved LTP.

Accordingly, the evaluation of PSR approval requests involving sites with groundwater or soil contamination may require more detailed technical analysis and additional technical assistance. The PM and technical reviewers shall contact the Reactor Decommissioning Section in the NMSS Division of Waste Management and Environmental Protection (NMSS/DWMEP) in addressing the implementation of the MOU at a specific reactor site. In addition, NUREG-1757 provides supporting guidance that may be used in evaluating such sites. (See Vol. 1, Section 9.3, and Vol. 2, Appendices F, G.2, H, and K.)

For PSRs involving radiologically impacted areas, a determination shall be made as to whether the site is characterized by unique conditions. This distinction is used to differentiate two types of sites: (1) sites that require only screening or simple analysis, and (2) sites that require detailed technical analysis and the use of advanced environmental models in assessing radiological consequences to offsite receptors. Sites with unique conditions generically may be characterized by one or more of the following complex features:

- radiological source-terms

- radiological release mechanisms

- surface and groundwater transport processes

- source term-to-receptor transport mechanisms

Such sites may have some of these features, but not necessarily all of them at a single site or a specific location on a site. Accordingly, the evaluation of PSR approval requests involving sites with unique conditions requires more detailed technical analysis and additional technical assistance. This approach is used to ensure that problematic technical issues are identified and resolved in a consistent manner. The PM shall contact the Reactor Decommissioning Section in NMSS/DWMEP to define the scope of the review and technical assistance. NUREG-1757 provides specific information and guidance that may be used in evaluating such sites (see Vol. 2, Section 1.3 and App. F, G, H, and I).

1.1.4   Site Visits and Inspections

Site visits and inspections may be conducted to give the PM and reviewers an opportunity to obtain first-hand knowledge of the site and its conditions. This phase of the review may be conducted in conjunction with the detailed technical review. A site inspection could be conducted to confirm the site conditions against their representation in the application package. Also, the inspection may include a review of records and documents (e.g., environmental monitoring data, HSA, etc.) that are referenced but not included in the application package. Site inspections may be performed by NRC regional office inspectors in conjunction with the review of a licensee's PSR request. Such inspections could include observations by inspectors, interviews with facility personnel (employees and contractors), selective examinations of records supporting the licensee's HSA, and an evaluation of radiological survey results generated by the licensee.

The inspection plan may be developed by the regional office in conjunction with the PM. The cognizant NRC regional office could have overall responsibility for conducting and reporting on the inspection, and may provide comments on the need and development of a confirmatory survey plan (see the related discussion below). The PM is responsible for providing technical support to the NRC regional office in developing the inspection plan, providing staff from NRC headquarters to assist the regional inspector, and providing information (inspection feeder) for inclusion in the inspection report prepared by the regional inspector. The PM and NRC regional office are responsible for coordinating the roles and participation of other agencies in such inspections.

1.1.5   Confirmatory Radiological Surveys

The PM is responsible for determining the need to conduct NRC confirmatory radiological surveys, following the evaluation of information provided by the licensee. The decision to conduct surveys will depend on (1) whether the area slated for PSR is designated as an impacted area, (2) whether the licensee has submitted its own survey results for a non-impacted area, and (3) the potential threat to public health and safety. Confirmatory surveys shall be conducted by NRC regional office inspectors or NRC contractors in conjunction with the site inspection, as outlined above.

Confirmatory radiological surveys may be performed (1) for the sole purpose of comparing specific NRC survey results with those of the licensee, or (2) to obtain an independent NRC assessment of the radiological conditions of the area, regardless of the results reported by the licensee. The surveys may be conducted during the detailed technical review phase, or possibly in parallel with the licensee's surveys. A confirmatory radiological survey of non-impacted areas is normally not needed if the staff agrees with the licensee's assessment that the area is non-impacted. However, a confirmatory survey of a non-impacted area may be conducted if the staff has unresolved questions concerning the licensee's assessment. In such cases, the decision to conduct a survey of a non-impacted area shall be identified early in the review process, as the assessment of survey results is likely to be a critical path item in the evaluation process. The survey plan shall be developed using the MARSSIM guidance (NUREG-1575) and NMSS consolidated guidance on decommissioning (NUREG-1757, Vol. 1, Sec. 15.4.5, and Vol. 2, Sec. 4).

The conduct of confirmatory radiological surveys shall be integrated in the site inspection, as outlined above. The cognizant NRC regional office will have the responsibility for this aspect of the inspection, and shall provide comments on the scope of the confirmatory survey plan. The cognizant PM from NRC headquarters is responsible for providing technical support to the regional office by defining survey objectives and criteria, providing staff to assist the regional inspector in conducting surveys, and providing information (inspection feeder) for inclusion in the inspection report prepared by the regional inspector. The PM and regional office are responsible for coordinating the participation of State and local agencies during such inspections (e.g., parallel survey measurements, split sample collection and analysis, etc.).

1.1.6 Consultations with Licensees

Before finalizing a PSR application, the NRC encourages licensees to meet with the PM to discuss the proposed PSR and address specific questions before submitting the application. For example, the discussion might address the basis for classifying an area as non-impacted or impacted, objectives and design of radiological surveys, and development of unrestricted release criteria. The staff should take this opportunity to identify appropriate and relevant guidance, given the specifics of the application. The SRP identifies prerequisites and refers to more exhaustive NRC guidance. The documents listed under "Regulatory Guidance" identify sections where specific information may be found. All consultations shall be arranged through the PM. The meeting purpose, location, date, and time shall be announced on the NRC's public Web site as a public meeting. Consultations with licensees may be included as part of site inspection activities.

1.1.7 Consultations with Other Agencies

The processing of an application may require the PM and reviewer(s) to contact other Federal, State, local, and regional agencies, and Native American tribal agencies. The NRC has issued a policy statement that addresses the participation of State regulatory agencies. The policy statement is entitled: "Cooperation with States at Commercial Nuclear Production or Utilization Facilities" (*Federal Register*, Vol. 54, p.7530, Feb. 22, 1989, as amended in Vol. 57, p.6462, Feb. 25, 1992). In addition, the procedures of 10 CFR 50.91 shall be considered whenever State consultations are mandated. All meetings shall be arranged through the PM. The meeting purpose, location, date, and time shall be announced on the NRC's public Web site. Meetings with such agencies may be included as part of site inspection activities.

Whenever groundwater or soil is known or suspected to be contaminated with plant-derived radioactivity, the PM shall consider the provisions of the MOU between the EPA and the NRC to identify their respective roles for decommissioning NRC-licensed sites (*Federal Register*, Vol. 67, No. 206, p.65375, Oct. 24, 2002). The MOU states that the EPA will defer its authority under CERCLA for the majority of facilities decommissioned under NRC authority. The MOU includes provisions for joint consultations for sites when, at the time of license termination, (1) groundwater contamination exceeds EPA-permitted levels, (2) the NRC contemplates restricted release or alternative criteria for the site, or (3) residual radioactive soil concentrations exceed levels defined in the MOU. The PM shall contact the Reactor Decommissioning Section in NMSS/DWMEP to address the implementation of the MOU at a specific reactor site and determine the need to consult with the EPA.

The PM and reviewer(s) may contact other Federal and State and local regulatory agencies in addressing requirements and compliance with environmental and health protection regulations governing the presence of toxic or hazardous properties of materials, if present in areas proposed for PSR. The presence of toxic or hazardous materials may be associated with spills and leaks involving the use of industrial chemicals, reagents, lubricants, and so forth.

1.1.8 Partial Site Release Evaluation Findings

The PM is responsible for coordinating the review and approval process with NRC offices and divisions to ensure that the requirements of 10 CFR 50.83 are met. The information provided by the licensee should demonstrate that there will be no adverse impact on the scope and conduct of programs associated with the remaining licensed site and facility operations as a result of the proposed PSR. The information and results of analyses should present an assessment of the impact on changing the site boundary and whether the PSR will result in changes to plant structures, systems, and components described in the FSAR or USAR. The application should identify and discuss potential changes in plant operations and technical specifications, changes in the radiological environmental monitoring program (REMP), changes in the offsite dose calculation manual (ODCM), changes in emergency planning and site security, changes in site criteria, and impacts on the operation of the independent spent fuel storage installation (ISFSI).

If the application is found to be incomplete or if deficiencies are identified during the review, the reviewer is responsible for requesting additional information or seeking clarification from the licensee in writing through the PM. The reviewer is responsible for documenting the findings of the review as mandated by NRC office or division procedures. A copy of the document shall be placed in the licensee's docket and in ADAMS.

If the NRC determines that the application is complete, technically acceptable, and meets the requirements of the regulations, the approval of a PSR request shall be documented by a license amendment or letter approval, as warranted. The approval process for a license amendment related to a PSR shall follow the procedures of 10 CFR 50.92 and NRC office and division procedures. The concurrence process for a letter documenting the approval of a PSR shall be the same as that for routine license amendments.

1.1.9 Approval Process

The approval of a PSR application via a letter or license amendment shall be supported by an SER documenting the technical review, evaluation, and the basis of the findings and acceptance. For a non-impacted area, the PM is responsible for informing the applicant (by letter) that the release is approved, and for preparing and issuing an EA. When a license amendment is required to implement a PSR, the PM is responsible for following the procedures of 10 CFR 50.92, and for preparing and issuing either an EA or an EIS.

The SER shall summarize the staff's evaluation and findings, and reference the inspection report prepared by the regional inspector. Technical and regulatory topics not considered in the review should be indicated in the SER, with reasons given for their exclusion. The findings noted in the inspection report shall be clearly stated in the SER. The preparation of the draft and final SER shall be initiated by the PM and coordinated with the cognizant regional office.

1.2 References

Code of Federal Regulations, Title 10, Section 2.206, "Requests for Action Under this Subpart."

Code of Federal Regulations, Title 10, Part 20, "Standards for Protection Against Radiation."

Code of Federal Regulations, Title 10, Section 20.1402, "Radiological Criteria for Unrestricted Use."

Code of Federal Regulations, Title 10, Part 50, "Domestic Licensing of Production and Utilization Facilities."

Code of Federal Regulations, Title 10, Section 50.2, "Definitions."

Code of Federal Regulations, Title 10, Section 50.75(g), "Reporting and Recordkeeping for Decommissioning Planning."

Code of Federal Regulations, Title 10, Section 50.83, "Release of Part of a Power Reactor Facility or Site for Unrestricted Use."

Code of Federal Regulations, Title 10, Section 51.53, "Post-Construction Environmental Reports."

Code of Federal Regulations, Title 10, Section 50.91, "Notice for Public Comment; State Consultation."

Code of Federal Regulations, Title 10, Section 50.92, "Issuance of Amendment."

Code of Federal Regulations, Title 10, Part 72," Licensing Requirements for the Independent Storage of Spent Nuclear Fuel, High-Level Radioactive Waste, and Reactor-Related Greater Than Class C Waste."

Code of Federal Regulations, Title 10, Part 100, "Reactor Site Criteria."

U.S. Nuclear Regulatory Commission, NUREG-1555, "Standard Review Plans for Environmental Reviews for Nuclear Power Plants, Environmental Standard Review Plan," Washington, DC, October 1999.

U.S. Nuclear Regulatory Commission, NUREG-1575, "Multi-Agency Radiation Survey and Site Investigation Manual (MARSSIM)," Rev. 1, Washington, DC, August 2000.

U.S. Nuclear Regulatory Commission, NUREG-1576, "Multi-Agency Radiological Laboratory Analytical Protocols Manual (MARLAP)," Washington, DC, July 2004.

U.S. Nuclear Regulatory Commission, NUREG-1700, "Standard Review Plan for Evaluating Nuclear Power Reactor License Termination Plans," Rev. 1, Washington, DC, April 2003a.

U.S. Nuclear Regulatory Commission, NUREG-1757, "Consolidated NMSS Decommissioning Guidance, Characterization, Survey, and Determination of Radiological Criteria," Vol. 1 & 2, Final Report, Washington, DC, September 2003b.

2. PARTIAL SITE RELEASE STANDARD REVIEW PLAN

This section presents technical guidance and regulatory requirements addressing the implementation of the regulations for PSRs. The following summarizes the regulatory framework that a licensee or applicant shall use to submit a request for a PSR, and steps the NRC staff shall use to review, evaluate, and process such a request:

- For non-impacted areas:

    - The applicant shall comply with 10 CFR 50.83(a)(1) and (a)(2) and 10 CFR 50.83(b)(1) to (b)(5).

    - The NRC review, evaluation, and processing steps are defined in 10 CFR 50.83(c) and 10 CFR 50.83(f).

    - The applicant shall comply with applicable parts of the SRP (Sections II.1 to II.4, II.7, and II.8).

- For impacted areas:

    - The applicant shall comply with 10 CFR 50.83(a)(1) – (a)(3) and 10 CFR 50.83(d)(1) – (d)(3).

    - The NRC review, evaluation, and processing steps are defined in 10 CFR 50.83(e) and 10 CFR 50.83(f).

    - The applicant shall comply with applicable parts of the SRP (Sections II.1 – II.3 and II.5 – II.8).

The specifics of each application will dictate which functional NRC office (division and branch) and region, and technical disciplines are needed to conduct the review and evaluation. The cognizant NRC PM shall identify primary and secondary responsibilities. The assignment of review responsibilities among primary and secondary reviewers is addressed in the previous section.

Each section of the SRP identifies areas of review; acceptance criteria; regulatory requirements and guidance; information to be submitted by the licensee; evaluation findings; and references. The PM and reviewers may select and emphasize particular elements of the SRP and define the corresponding level of technical review. In such instances, the staff may not carry out, in detail, all of the review steps described in each section of the SRP.

Reviewers are responsible for documenting all findings, discrepancies, missing information, etc., as mandated by office and division procedures. A copy of the document shall be placed in the licensee's docket and in ADAMS. When deficiencies are identified, the PM shall request the additional information or seek clarification from the licensee in writing.

2.1 General Information and Description of Area Proposed for Partial Site Release

The regulations require the licensee to provide information, such as maps, plot plans, etc., describing the portion of the site to be released, and define its location and perimeter using a site survey grid coordinate system. The information should include a complete description of all major site features; location of the proposed area in relation to the property defined in the facility license; description of surrounding offsite environs and resources; planned or tentative schedule for the effective date of the proposed release; and whether the area to be released includes structures, facilities, and utilities supporting site operations. The required information is described in 10 CFR 50.83(b)(1) – (b)(3) for non-impacted areas, and 10 CFR 50.83(d)(1) for impacted areas.

2.1.1 Areas of Review

The information provided by the licensee should be as complete as is possible to permit an assessment of the relationship of the area to be released with potential impacts on site environs and resources. The information should include sufficient details about the physical characteristics of the area to be released, as well as its relationship to the remaining licensed site and surrounding areas. This information should be included on updated maps and site plot plans with sufficient details showing all relevant features, along with supporting text for details that cannot be readily included on site maps or plans. The maps may be based or rely, in part, on information contained in existing reports, such as an environmental report, an FSAR or USAR, or the ODCM.

2.1.2 Review Procedures

2.1.2.1 Acceptance Review

The NRC staff shall review and confirm that the application contains the information summarized under "Areas of Reviews," and described under "Information to be Submitted." The acceptance review shall confirm that the required information is complete to conduct a detailed technical analysis as the next step. The NRC staff shall confirm, via a preliminary review of the table of contents, content of each section, and any supporting attachments, that the licensee has provided the required information and determine that its level of detail is adequate. If the staff determines that the information is incomplete or insufficient to support a detailed technical analysis, the staff shall compile a list of such deficiencies and describe why the information is inadequate.

2.1.2.2 Safety Evaluation

The material to be provided by the licensee covers specific technical information about the site and facility features, and, consequently, the staff is responsible for reviewing and evaluating the information to a depth that is commensurate with the level of technical detail. The staff shall confirm that the licensee has provided appropriate information and the site and facility descriptions are current and valid. The staff shall also confirm that the licensee has provided the required information to allow independent confirmation of the status of the site and facilities, and area(s) to be released under 10 CFR 50.83(a) and (b) for non-impacted areas, or 10 CFR 50.83(a) and (d) for impacted areas, and that the staff has followed the review and evaluation process of 10 CFR 50.83(c) and (f) for non-impacted areas, or 10 CFR 50.83(e) and (f) for impacted areas.

2.1.3 Acceptance Criteria

2.1.3.1 Regulatory Requirements

- 10 CFR 20.101, "Filing of Application."
- 10 CFR 50.83(c) and (e), "Release of Part of a Power Reactor Facility or Site for Unrestricted Use."

2.1.3.2 Regulatory Guidance

- NUREG-1757, Vol. 1, "Consolidated NMSS Decommissioning Guidance":
  - Section 16, "Decommissioning Plans: Site Description."

- NUREG-1555, "Standard Review Plans for Environmental Reviews for Nuclear Power Plants":
  - Section 2, "Environmental Description."
  - Section 3, "Plant Description."
  - Section 5, "Environmental Impacts of Station Operation."

2.1.3.3 Information To Be Submitted

The reviewer shall verify that the licensee has provided the following information regarding the area to be released:

- the licensee's name and address, license number, and docket number, as well as the location of the site if physically different than the licensee's address

- the proposed schedule for the release of the property and its expected future use

- the size and location of the area to be released, including its proximity to the remainder of the site

- a description of facilities in the area to be released, including buildings, roads; parking lots; fixed equipment; storage tanks; settling ponds; surface streams; storage facilities for hazardous and non-hazardous materials; disposal areas for industrial, construction, and demolition wastes; underground and overhead utilities; wells; and facilities supporting site operations, among other features

- a description of contour elevations, site elevation, and natural features of the area to be released

- a description of any changes in site features, such as construction, demolition, surface regrading, and surface or groundwater hydrology, as appropriate

- an assessment of potential surface and subsurface mechanisms for transporting radioactive contaminants from the licensed portions of the site to the area to be released

Using information from prior licensee submissions (e.g., environmental report(s), FSARs or USARs, ODCM, etc.), the reviewer shall verify that the licensee has properly characterized changes to or impacts on nearby areas that may be affected by the PSR, including the following updated information:

- descriptions of properties surrounding the area to be released, such as nearby residences, businesses, and community facilities

- characterizations of nearby communities, towns, cities, and Native American tribal areas

- information on nearby prominent features, such as highways, roads, streets, rails, local landmarks, parks, rivers, and lakes

2.1.4 Evaluation Criteria and Findings

Reviews performed for this section of the SRP are based on guidance and criteria listed under "Regulatory Requirements" and "Regulatory Guidance," above. The reviewer shall verify that, where applicable, the licensee's application includes the information summarized under "Information to be Submitted," above. The staff's review shall verify, to the maximum extent practicable, that the information supplied by the licensee is complete and accurate by comparing it with prior licensee submissions, prior environmental report(s), FSARs or USARs, OCDM, licensing actions, and inspection records maintained in NRC files. The reviewer shall verify that the information is provided and presented in a manner consistent with the sections of NUREG-1555 and NUREG-1757 referenced above.

2.1.5 References

U.S. Nuclear Regulatory Commission, NUREG-1555, "Standard Review Plans for Environmental Reviews for Nuclear Power Plants," Washington, DC, October 1999.

U.S. Nuclear Regulatory Commission, NUREG-1757, "Consolidated NMSS Decommissioning Guidance, Decommissioning Process for Materials Licensees," Vol. 1, Final Report, Washington, DC, September 2003.

2.2 Characterization of Area Proposed for Partial Site Release

The prior and current uses of the area proposed for partial site release should be known for the purpose of characterizing its current radiological status. The characterization is based on the results of an HSA, results of past and current radiation surveys, process knowledge, and, if needed, conduct of new characterization surveys. The information is used to determine whether the area is radiologically impacted or non-impacted. Under 10 CFR 50.2, the definitions of "impacted areas" and "non-impacted areas" are as follows:

- "*Impacted areas* mean the areas with some reasonable potential for residual radioactivity in excess of natural background or fallout levels."

- "*Non-impacted areas* mean the areas with no reasonable potential for residual radioactivity in excess of natural background or fallout levels."

Additional technical details are provided in NUREG-1575 and NUREG-1757. For non-impacted areas, the results of the HSA and any supplemental source of information are used to document the technical basis and justification for declaring the area non-impacted. For an area defined as radiologically impacted, the information is used to plan and implement additional radiation surveys, assess radioactivity levels, assign the appropriate MARSSIM survey classification to the area, and assess the need for remediation to meet release criteria.

The required information is described in 10 CFR 50.83(a) and (b) for non-impacted areas, and 10 CFR 50.83(a) and (d) for impacted areas. For a PSR involving an impacted area, this information shall be used by the licensee to prepare a license amendment request and a supplement to the facility's environmental report under 10 CFR 50.83(d)(3).

2.2.1 Areas of Review

The reviewer is responsible for confirming that the submitted information provides the means to determine the extent and levels of radioactivity and radionuclide distributions within the area to be released. This information, as appropriate, should address facility grounds, structures, and foundations; systems and components remaining in the area after release; residues, if any, in process system components; area soils (surface and subsurface); site utilities, such as effluent outfalls, sewers, and surface site drainage structures; and surface and groundwater resources. The licensee's description and characterization of the area should be comprehensive to allow the staff to conclude, using MARSSIM guidance, that the HSA and supplemental information support the appropriate radiological designation for the area as impacted or non-impacted, as defined in 10 CFR 50.2.

2.2.2 Review Procedures

2.2.2.1 Acceptance Review

The NRC staff shall review and confirm that the application contains the information summarized under "Areas of Reviews," and described under "Information to be Submitted." The acceptance review shall confirm that the required information is complete to conduct a detailed technical analysis as the next step. The NRC staff shall confirm, via a preliminary review of the table of contents, content of each section, and any supporting attachments, that the licensee has provided the required information and determine that its level of detail is adequate. If the staff determines that the information is incomplete or insufficient to support a detailed technical analysis, the staff shall compile a list of such deficiencies and describe why the information is inadequate.

2.2.2.2 Safety Evaluation

The material provided by the licensee addresses specific technical information about the area(s) to be released. The staff shall review and evaluate the information at a level commensurate with the level of technical detail. The staff shall confirm that the licensee has used current site information and the characterization of the area proposed for PSR reflects its current radiological conditions. The staff shall confirm that the licensee has provided the required information to allow independent confirmation of the conditions of the area to be released under 10 CFR 50.83(a) and (b) for non-impacted areas, or 10 CFR 50.83(a) and (d) for impacted areas, and that the staff has followed the review and evaluation process of 10 CFR 50.83(c) and (f) for non-impacted areas, or 10 CFR 50.83(e) and (f) for impacted areas.

2.2.3 Acceptance Criteria

2.2.3.1 Regulatory Requirements

- 10 CFR 50.2, "Definitions."
- 10 CFR 50.83, "Release of Part of a Power Reactor Facility or Site for Unrestricted Use."
- 10 CFR 20.1402, "Radiological Criteria for Unrestricted Use."

Note: The disposition of radioactive materials or waste in an area designated for a PSR under the provisions of the following regulations should be included in the site description as applicable.

- 10 CFR 20.2002, "Method for Obtaining Approval of Proposed Disposal Procedures" (and its former counterpart under 10 CFR 20.302).
- 10 CFR 20.2003, "Disposal by Release Into Sanitary Sewerage."
- 10 CFR 20.2004, "Treatment or Disposal by Incineration."
- 10 CFR 20.304, "Disposal by Burial in Soil" (for practices conducted before its rescission on October 30, 1980).

2.2.3.2 Regulatory Guidance

- NUREG-1575, Rev. 1, "Multi-Agency Radiation Survey and Site Investigation Manual (MARSSIM)":
  - Section 3.0, "Historical Site Assessment."
  - Section 4.4, "Classify Areas by Contamination Potential."
  - Section 4.9, "Quality Control."
  - Section 5.3, "Characterization Surveys."
  - Section 7.0, "Sampling and Preparation for Laboratory Measurements."
  - Section 9.0, "Quality Assurance and Quality Control."

- NUREG-1576, Vol. 2, "Multi-Agency Radiological Laboratory Analytical Protocols Manual (MARLAP)":
  - Section 10, "Field Sampling Issues That Affect Laboratory Measurements."

- NUREG-1757, Vol. 2, "Consolidated NMSS Decommissioning Guidance":
  - Section 4.2, "Characterization Surveys."
  - Section 5.0, "Dose Modeling Evaluation."
  - Appendix A.1, "Classification of Areas by Residual Radioactivity Levels."
  - Appendix A.5, "Instrument Selection and Calibration."
  - Appendix D, "Survey Data Quality and Reporting."
  - Appendix E, "Measurements for Facility Radiation Surveys."
  - Appendix F, "Ground and Surface Water Characterization."
  - Appendix G, "Special Characterization and Survey Issues."
  - Appendix H, "Criteria for Conducting Screening Dose Modeling Evaluations."
  - Appendix I, "Technical Basis for Site-Specific Dose Modeling Evaluations."
  - Appendix K, "Dose Modeling Considerations for Partial Site Release."
  - Appendix N, "ALARA Analyses."

2.2.3.3 Information To Be Submitted

For areas designated as radiologically impacted, the reviewer shall verify that the licensee has provided the following information:

- basis for the designation of radiologically "impacted," as founded on the results of a site-specific HSA

- locations and descriptions of radiological discharges, disposals, spills, unmonitored releases, operational activities, or radiological accidents/incidents that have occurred and resulted in known or suspected contamination of the site or area to be released

- radiological characterization results of the area to be released, where radioactivity is known or suspected to be present, including radionuclide distributions, maximum and average radionuclide concentrations, and ambient external radiation exposure rates, including the following information:

  - description of the extent and levels of radioactivity in surface and subsurface soils and groundwater.

  - description of the extent and levels of radioactivity in structures, buildings, foundations, and site features, such as underground utilities, paved areas (parking areas and walkways), surface drainage structures, settling or water storage ponds, site surface drainage outfalls, etc.

  - description of the extent and levels of radioactivity in process systems, such as components, floor drains, ventilation ducts, piping and embedded piping, sewers, service utilities, etc.

  - description of the extent and levels of radioactivity within waste management facilities, such as process systems and components, process residues, material storage or staging areas, demolition debris, waste disposal and storage areas, effluent discharge points, etc.

- description of past and recent remediation activities conducted within the area to be released and its immediate vicinity, including supporting ALARA analyses, if necessary

- inventories of material and radioactivity levels associated with prior onsite disposals (made under 10 CFR 20.2002, 20.302, and 20.304), including radionuclide distributions and concentrations, and physical and chemical characterizations of such wastes

- inventories of materials and radioactivity levels disposed of by releases into sanitary sewerage or by incineration, including radionuclide distributions and concentrations, and physical and chemical characterizations of such wastes

- survey methods and instrumentation used in characterizing the area and supporting quality assurance/control measures used in conducting associated radiation surveys, including sample collection and analysis

- representation of ambient background radioactivity and radiation levels used or reported during the conduct of scoping or characterization surveys

- assessment of potential for re-contamination of the area, description of the origin and form of the radioactivity, process mechanisms that could result in re-contamination, and description of administrative and engineered controls established to prevent it

- description of any previous PSRs implemented at the licensed site, radiological designation (impacted or non-impacted) of the area at the time the release was granted, basis of release criteria, and documentation demonstrating compliance

- basis for the MARSSIM survey classification assigned to the area proposed for PSR

- supplemental site characterization information supporting the justification for site-specific exposure scenarios and pathways if the licensee proposes to derive site-specific concentration guideline levels in compliance with 10 CFR 20.1402

For areas designated as radiologically "non-impacted," the reviewer shall verify that the licensee has provided the following information:

- basis for the radiological designation of "non-impacted," as founded on the results of a site-specific HSA

- detailed summary of the HSA and rationale supporting the designation of "non-impacted," including supporting references and documentation, as needed

- if the licensee voluntarily conducted a radiological survey to confirm the results of its HSA providing the following information:

  - description of survey methods used to assess the radiological status of the area

  - results of quality assurance/control measures used in conducting the associated radiation surveys, including sample collection and analysis

  - discussion of the basis for assigned ambient background radioactivity and radiation levels in the area, or results reported during surveys to confirm the radiological status of the area

- assessment of potential for re-contamination of the area, description of the origin and form of the radioactivity, process mechanisms that could result in re-contamination, and description of administrative and engineered controls established to prevent it

- description of any previous PSR implemented at the licensed site, radiological designation (impacted or non-impacted) of the area at the time the release was granted, basis of release criteria, and documentation demonstrating compliance

2.2.4 Evaluation Criteria and Findings

Reviews performed for this section of the SRP are based on guidance and criteria listed under "Regulatory Requirements" and "Regulatory Guidance," above. The reviewer shall verify that, where applicable, the licensee's application includes the information summarized under "Information to be Submitted," above. The staff's review shall verify, to the maximum extent practicable, that the information supplied by the licensee is complete and accurate by comparing it with prior licensee submissions, licensing actions, and inspection records maintained in NRC files. The reviewer shall verify that the radiological characterization of the site is adequately described, the licensee's conclusions as to the radiological designation and survey classification of the area are supported by the HSA, appropriate resolutions of inconsistencies or information gaps in reconstructing the operational history of the site or facility are included, site characterization data are complete and current, and the processes described in the referenced sections of NUREG-1575, NUREG-1576, and NUREG-1757 have been considered and used, as applicable.

Whenever groundwater or soil is known or suspected to be contaminated with plant-derived radioactivity, the PM will consider the provisions of the MOU between the EPA and NRC to identify their respective roles for decommissioning NRC-licensed sites (*Federal Register*, Vol. 67, No. 206, p.65375, Oct. 24, 2002). The MOU includes provisions identifying the need for joint consultations for sites that, at the time of license termination, (1) have groundwater contamination in excess of EPA-permitted levels, (2) the NRC contemplates restricted release or alternative criteria for the site, or (3) have residual radioactive soil concentrations in excess of levels defined in the MOU. The evaluation of PSR requests for sites known or suspected to have contaminated groundwater or soil shall be evaluated on a case-specific basis. The PM and technical reviewers shall contact the Reactor Decommissioning Section in NMSS/DWMEP in addressing the implementation of the MOU at a specific reactor site and a determination shall be made as to whether the site is eligible for partial site release. In addition, NUREG-1757 provides supporting guidance that may be used in evaluating such sites (see Vol. 1 Section 9.3, and Vol. 2, Appendices F, G.2, H, and K).

For PSRs involving radiologically impacted areas, a determination will be made as to whether the site is characterized by unique conditions. This distinction is used to differentiate (1) sites that require only screening or simple analysis, and (2) sites that require detailed technical analysis and use of advanced environmental models in assessing radiological consequences to offsite receptors. Sites with unique conditions are characterized generically by one or more of the following complex features:

- radiological source-terms

- radiological release mechanisms

- surface and groundwater transport processes

- source term-to-receptor transport mechanisms

Such sites may have some of these features, but not necessarily all of them at a specific site or location on a site. Accordingly, the evaluation of PSR approval requests involving sites with unique conditions may require more detailed technical analysis and additional technical assistance. This approach is used to ensure that problematic technical issues are identified and resolved in a consistent manner. The PM shall contact the Reactor Decommissioning Section in NMSS/DWMEP to define the scope of the review and technical assistance. NUREG-1757 provides specific information and guidance that may be used in evaluating such sites (see Vol. 2, Section 1.3 and App. F, G, H, and I).

2.2.5 References

U.S. Nuclear Regulatory Commission, NUREG-1575, "Multi-Agency Radiation Survey and Site Investigation Manual (MARSSIM)," Rev. 1, Washington, DC, August 2000.

U.S. Nuclear Regulatory Commission, NUREG-1576, "Multi-Agency Radiological Laboratory Analytical Protocols Manual (MARLAP)," Washington, DC, July 2004.

U.S. Nuclear Regulatory Commission, NUREG-1757, "Consolidated NMSS Decommissioning Guidance, Characterization, Survey, and Determination of Radiological Criteria," Vol. 2, Final Report, Washington, DC, September 2003.

2.3   License Program Impacts

In assessing impacts, the licensee must evaluate how the PSR will affect potential doses to members of the public from site operations, radiological effluents, emergency planning, site security, environmental monitoring, and site criteria of 10 CFR Part 100. These requirements are addressed in 10 CFR 50.83(a)(1), (b)(4), and (b)(5) for non-impacted areas, and 10 CFR 50.83(a)(1), (d)(1), and (d)(3) for impacted areas. As part of the conditions of the operating license, licensees shall provide assurances that the facility will be operated safely, and that public health and safety shall be maintained following any changes to the facility or site boundary. The bases and conclusions supporting the continued safe operation of the facility are contained in the FSAR or USAR prepared under the requirements of 10 CFR 50.71.

2.3.1 Areas of Review

The information provided by the licensee shall demonstrate that there will be no adverse impacts on the facility's license program associated with the remaining licensed site and facility operations as a result of the proposed PSR. The application should provide the results of evaluations performed under 10 CFR 50.59, and should include any relevant supporting information. The information and results of analysis should present an assessment of the impact of changing the site boundary and whether the PSR will result in changes to plant structures, systems, and/or components described in the FSAR or USAR. Moreover, the application should identify and discuss potential changes in plant operations and technical specifications, changes in the REMP and/or ODCM, changes in emergency planning and site security, changes in site criteria, and effects on the operation of the ISFSI, if present.

If the proposed PSR is expected to affect ongoing activities of the REMP, the licensee should address changes to the program in accordance with Appendix I to 10 CFR Part 50, and should incorporate the necessary changes in a revised ODCM. The scope of the REMP should reflect the guidance of Regulatory Guides 4.1, 4.8, and 4.15; NUREG-0800 (Section 11.5); and the Radiological Assessment Branch Technical Position (BTP), taking into account site-specific characteristics. If it is expected that changes associated with the PSR might result in deletion of some sampling and measurement locations, the licensee is expected to replace the deleted sampling locations by identifying new locations, thereby preserving the original design objective of the REMP. The licensee is expected to identify all new sampling locations, sampling media, and sample collection frequencies, and justify the reasons for additional or modified radiological analysis for such samples. If the PSR is not expected to result in changes in the REMP, the current scope and implementation of the REMP shall remain as is.

The proposed PSR might affect plant technical specifications and requirements of the ODCM that are used in demonstrating compliance with the effluent concentrations specified in Appendix B to 10 CFR Part 20, dose objectives specified in Appendix I to 10 CFR Part 50, and the fuel cycle dose limit specified in 40 CFR Part 190. As a result, plant technical specifications and methods presented in the ODCM may need to be updated to ensure that the accuracy or reliability of effluent release, dose, or set-point calculations are commensurate with changes associated with the PSR.

Some elements of the site emergency plan might also be impacted by the PSR. The licensee should address changes to the emergency plan in response to 10 CFR 50.54(q), 10 CFR 50.47, and Appendix E to 10 CFR Part 50. The evaluation should assess whether redefining the site boundary would have any impact on the effectiveness of the emergency response plan and implementing procedures, and/or compliance with the dose limits of 10 CFR Part 100 at the exclusion area boundary (EAB). Any potential impacts on ISFSI operations should also be evaluated and integrated in the emergency response plan and procedures.

The PSR might impact some elements of the site safeguard contingency plan and procedures. The licensee should address changes to the contingency plan and its procedures in response to 10 CFR 50.54(p), 10 CFR Part 73, and Appendix C to 10 CFR Part 73. The application should assess whether the resulting change to the site boundary might impact the effectiveness of the security plan, security system, implementing procedures, and other licensed portions of the site (i.e., ISFSI) under 10 CFR Part 72.

The proposed PSR might also affect site criteria requirements under 10 CFR Part 100. The application should assess whether the resulting change to the site boundary might impact the definition of the EAB and/or the area comprising the ISFSI, the determination of the low population zone (LPZ), and/or the commitment to meet the dose criteria of 10 CFR Part 100.

2.3.2 Review Procedures

2.3.2.1 Acceptance Review

The NRC staff shall review and confirm that the application contains the information summarized under "Areas of Reviews," and described under "Information to be Submitted." The acceptance review shall confirm that the required information is complete to conduct a detailed technical analysis as the next step. The NRC staff shall confirm, via a preliminary review of the table of contents, content of each section, and any supporting attachments, that the licensee has provided the required information and determine that its level of detail is adequate. If the staff determines that the information is incomplete or insufficient to support a detailed technical analysis, the staff shall compile a list of such deficiencies and describe why the information is inadequate.

2.3.2.2 Safety Evaluation

The material provided by the licensee addresses specific technical subjects regarding site and facility operations and how the licensee demonstrates compliance with its license and associated regulatory requirements. The staff shall review and evaluate the information at a level commensurate with the technical detail. The staff shall confirm that the licensee has used defensible information, and that descriptions of site and facility operations are current and can withstand scrutiny against technical and regulatory requirements. The staff shall confirm that the licensee has provided the required information to allow an independent assessment of the impacts of the PSR on site and facility operations under 10 CFR 50.83(a) and (b) for non-impacted areas, or 10 CFR 50.83(a) and (d) for impacted areas, and that the staff has followed the review and evaluation process of 10 CFR 50.83(c) and (f) for non-impacted areas, or 10 CFR 50.83(e) and (f) for impacted areas.

2.3.3 Acceptance Criteria

2.3.3.1 Regulatory Requirements

- 10 CFR Part 20, "Standards for Protection Against Radiation."

- 10 CFR Part 20, Subpart D, "Radiation Dose Limits for Individual Members of the Public."

- 10 CFR 20.1301, "Dose Limits for Individual Members of the Public"

- 10 CFR 50.36a, "Technical Specifications on Effluents from Nuclear Power Reactors."

- 10 CFR Part 50, "Domestic Licensing of Production and Utilization Facilities," Appendix I, "Numerical Guides for Design Objectives and Limiting Conditions for Operation to Meet the Criterion 'As Low As Is Reasonably Achievable' for Radioactive Material in Light-Water-Cooled Nuclear Power Reactor Effluents."

- 10 CFR Part 50, "Domestic Licensing of Production and Utilization Facilities," Appendix E, "Emergency Planning and Preparedness for Production and Utilization Facilities."

- 10 CFR 50.47, "Emergency Plans."

- 10 CFR 50.54, "Conditions of Licenses."

- 10 CFR 50.59(d)(1) to (d)(3), "Changes, Tests, and Experiments."

- 10 CFR 50.71, "Maintenance of Records, Making of Reports."

- 10 CFR 50.83, "Release of Part of a Power Reactor Facility or Site for Unrestricted Use."

- 10 CFR Part 72, "Licensing Requirements for the Independent Storage of Spent Nuclear Fuel, High-Level Radioactive Waste, and Reactor-Related Greater Than Class C Waste."

- 10 CFR Part 73, "Physical Protection of Plants and Materials."

- 10 CFR Part 73, "Physical Protection of Plants and Materials," Appendix C, "Licensee Safeguards Contingency Plans."

- 10 CFR Part 100, "Reactor Site Criteri.a"

- 40 CFR Part 190, "Environmental Radiation Protection Standards for Nuclear Power Operations."

2.3.3.2 Regulatory Guidance

- NUREG-0800, "Standard Review Plan," Section 11.5, "Process and Effluent Radiological Monitoring Instrumentation and Sampling Systems."

- NUREG-1301, "Offsite Dose Calculation Manual Guidance: Standard Radiological Effluent Controls for Pressurized Water Reactors."

- NUREG-1302, "Offsite Dose Calculation Manual Guidance: Standard Radiological Effluent Controls for Boiling Water Reactors."

- NUREG-1757, Vol. 2, "Consolidated NMSS Decommissioning Guidance":
  - Section 3.4, "Considerations for Other Constraints on Allowable Residual Radioactivity."
  - Appendix K, "Dose Modeling for Partial Site Release."

- Regulatory Issue Summary (RIS) 2000-19, "Partial Release of Reactor Site for Unrestricted Use Before NRC Approval of the License Termination Plan."

- Regulatory Guide 1.21, "Measuring, Evaluating, and Reporting Radioactivity in Solid Wastes and Releases of Radioactive Materials in Liquid and Gaseous Effluents from Light-Water-Cooled Nuclear Power Plants."

- Regulatory Guide 1.78, "Evaluating the Habitability of a Nuclear Power Plant Control Room During a Postulated Hazardous Chemical Release."

- Regulatory Guide 1.109, "Calculation of Annual Doses to Man from Routine Releases of Reactor Effluents for the Purpose of Evaluating Compliance with 10 CFR Part 50, Appendix I."

- Regulatory Guide 1.111, "Methods for Estimating Atmospheric Transport and Dispersion of Gaseous Effluents in Routine Releases from Light-Water-Cooled Reactors."

- Regulatory Guide 1.112, "Calculation of Releases of Radioactive Materials in Gaseous and Liquid Effluents from Light-Water-Cooled Power Reactors."

- Regulatory Guide 1.113, "Estimating Aquatic Dispersion of Effluents from Accidental and Routine Reactor Releases for the Purpose of Implementing Appendix I."

- Regulatory Guide 1.101, "Emergency Planning and Preparedness for Nuclear Power Reactors."

- Regulatory Guide 1.181, "Content of Updated Final Safety Analysis Report in Accordance with 10 CFR 50.71(e)."

- Regulatory Guide 4.1, "Programs for Monitoring Radioactivity in the Environs of Nuclear Power Plants."

- Regulatory Guide 4.8, "Environmental Technical Specifications for Nuclear Power Plants."

- Regulatory Guide 4.15, "Quality Assurance for Radiological Monitoring Programs (Normal Operations): Effluent Streams and the Environment."

- U.S. Nuclear Regulatory Commission, "Radiological Assessment Branch Technical Position (BTP)."

2.3.3.3 Information To Be Submitted

The reviewer shall verify that the licensee has provided sufficient information to demonstrate that the proposed PSR will not impede the following criteria:

- The residual dose from the PSR and the dose associated with site operations to individual members of the public will not exceed the limits of 10 CFR Part 20, Subpart D, including the requirements of 40 CFR Part 190.

- There will be no reduction in the effectiveness of emergency planning or physical security.

- Radiological effluent releases (water and air) will remain within the regulatory limits of 10 CFR Part 20 and dose objectives of Appendix I to 10 CFR Part 50.

- The dose criteria of 10 CFR Part 100 at the EAB and LPZ will continue to be met.

- There are no adverse impacts on other portions of licensed facilities, such as an ISFSI, as a result of changes to the site boundary.

- To the extent that the future use of the released property is known, an assessment of the presence of potentially hazardous facilities or activities involving hazardous materials that may affect facility operations and control room habitability.

- The results of 10 CFR 50.59 analyses and supporting information, as needed.

- Revisions made to radiological effluent technical specifications and limiting conditions for operation are documented in a revised ODCM.

- Confirmation that the REMP remains in compliance with Appendix I to 10 CFR Part 50; the guidance of Regulatory Guides 4.1, 4.8, and 4.15; and the Radiological Assessment BTP, taking into account site-specific characteristics.

- Description and justification of all changes to the REMP and ODCM, commensurate with the impact associated with the PSR, including changes in sampling locations, environmental media sample, collection frequencies, and the types and frequencies of radiological analysis.

- Description of revisions made to the emergency site plan and implementing procedures, as needed.

- Description of revisions made to site physical security features and operation of the site security system, as needed.

- Commitment that all other applicable statutory and regulatory requirements of 10 CFR Parts 20, 50, 51, 72, and 100, which may be affected as a result of the release or changes to the site boundary, will continue to be met as specified in the license and FSAR or USAR.

2.3.4 Evaluation Criteria and Findings

Reviews performed forn this section of the SRP are based on guidance and criteria listed under "Regulatory Requirements" and "Regulatory Guidance," above. The reviewer shall verify that, where applicable, the licensee's application includes the information summarized under "Information to be Submitted," above. The staff's review shall verify, to the maximum extent practicable, that the information supplied by the licensee is complete and accurate by comparing it with prior licensee submissions, FSAR or USAR, licensing actions, and inspection records maintained in NRC files. The review shall confirm that the partial site release will not (1) result in exceeding dose limits to members of the public from all aspects of site and facility operations, and/or (2) reduce the effectiveness of the environmental monitoring program, site criteria, emergency planning and response, and/or site security measures.

In the case of a release involving radiologically impacted property, the licensee's demonstration of compliance with the public dose limits and standards requires a discussion of compliance with the EPA's fuel cycle radiation standard (40 CFR Part 190) incorporated in 10 CFR 20.1301(e). NUREG-1757 (Vol. 2, Sec. 3.4 and App. K.2) provides additional guidance on demonstrating continued compliance with the EPA dose standards in 40 CFR Part 190 when releasing radiologically impacted property for unrestricted use.

The staff shall assess whether the proposed PSR might adversely affect plant technical specifications and requirements of the ODCM used in demonstrating compliance with the effluent concentrations specified in Appendix B to 10 CFR Part 20, dose objectives specified in Appendix I to 10 CFR Part 50, and the fuel cycle dose limit specified in 40 CFR Part 190. If the plant technical specifications and methods used in the ODCM are impacted, the staff shall confirm that the required changes to technical specifications and ODCM methods will not degrade the accuracy or reliability of quantifying effluent release rates, calculating doses, or determining alarm set-points of instruments monitoring effluent releases.

The staff shall assess whether the proposed PSR might adversely affect some elements of the REMP, and, if so, confirm that the required changes to the program conform with Appendix I to 10 CFR Part 50 and the ODCM. If the REMP is impacted, the staff shall also confirm that deleted sampling and monitoring locations have been replaced by new ones, thereby preserving the original design objective of the REMP. The staff shall also assess the adequacy of all new sampling locations, sampling media, sample collection frequencies, and stated justification for the additional or modified radiological analysis for such samples, if needed.

The staff shall assess whether the site emergency plan is adversely affected by the proposed PSR. If so, the staff shall evaluate changes to the emergency plan in response to 10 CFR 50.54(q), 10 CFR 50.47, and Appendix E to 10 CFR Part 50. The evaluation shall assess whether

(1) the licensee is taking credit for the redefinition of the site boundary in the emergency plan, (2) the new site boundary line might have an impact on the effectiveness of the emergency response plan and implementing procedures, and (3) the licensee remains in compliance with the dose limits of 10 CFR Part 100 at the EAB, including the ISFSI.

The staff shall assess whether the PSR might adversely affect the site safeguard contingency plan. The evaluation shall consider changes to the contingency plan and procedures in response to 10 CFR 50.54(p), 10 CFR Part 73, and Appendix C to 10 CFR Part 73. The staff shall determine whether the licensee is taking credit for the redefinition of the site boundary in the security plan, and assess the effectiveness of the security plan, security system, and implementing procedures, as well as any changes of security measures for the ISFSI.

The staff shall assess whether the proposed PSR might adversely affect site factors and criteria requirements. The evaluation shall assess whether the resulting change in the site boundary might adversely affect site criteria, description of the EAB and/or LPZ, and commitment to meet the dose criteria of 10 CFR Part 100 at the EAB and LPZ.

2.3.5   References

U.S. Nuclear Regulatory Commission, Regulatory Issue Summary (RIS) 2000-19, "Partial Release of Reactor Site for Unrestricted Use Before NRC Approval of the License Termination Plan," Washington, DC, October 24, 2000.

U.S. Nuclear Regulatory Commission, NUREG-1757, "Consolidated NMSS Decommissioning Guidance, Characterization, Survey, and Determination of Radiological Criteria," Vol. 2, Final Report, Washington, DC, September 2003.

U.S. Nuclear Regulatory Commission, NUREG-0800, "Standard Review Plan," Section 11.5, "Process and Effluent Radiological Monitoring Instrumentation and Sampling Systems," Draft Rev.4, Washington, DC, April 1996.

U.S. Nuclear Regulatory Commission, NUREG-1301, "Offsite Dose Calculation Manual Guidance: Standard Radiological Effluent Controls for Pressurized Water Reactors," Washington, DC, April 1991.

U.S. Nuclear Regulatory Commission, NUREG-1302, "Offsite Dose Calculation Manual Guidance: Standard Radiological Effluent Controls for Boiling Water Reactors," Washington, DC, April 1991.

U.S. Nuclear Regulatory Commission, Washington DC, Regulatory Guides:

- Regulatory Guide 1.21, "Measuring, Evaluating, and Reporting Radioactivity in Solid Wastes and Releases of Radioactive Materials in Liquid and Gaseous Effluents from Light-Water-Cooled Nuclear Power Plants," Rev. 1, June 1974.

- Regulatory Guide 1.78, "Evaluating the Habitability of a Nuclear Power Plant Control Room During a Postulated Hazardous Chemical Release," Rev. 1, December 2001.

- Regulatory Guide 1.109, "Calculation of Annual Doses to Man from Routine Releases of Reactor Effluents for the Purpose of Evaluating Compliance with 10 CFR Part 50, Appendix I," Rev. 1, October 1977.

- Regulatory Guide 1.111, "Methods for Estimating Atmospheric Transport and Dispersion of Gaseous Effluents in Routine Releases from Light-Water-Cooled Reactors," Rev. 1, July 1977.

- Regulatory Guide 1.112, "Calculation of Releases of Radioactive Materials in Gaseous and Liquid Effluents from Light-Water-Cooled Power Reactors," May 1977.

- Regulatory Guide 1.113, "Estimating Aquatic Dispersion of Effluents from Accidental and Routine Reactor Releases for the Purpose of Implementing Appendix I," Rev. 1, April 1977.

- Regulatory Guide 1.101, "Emergency Planning and Preparedness for Nuclear Power Reactors," Rev. 4, July 2003.

- Regulatory Guide 1.181, "Content of Updated Final Safety Analysis Report in Accordance with 10 CFR 50.71(e)," Rev. 1, March 1999.

- Regulatory Guide 4.1, "Programs for Monitoring Radioactivity in the Environs of Nuclear Power Plants," Rev.1, April 1975.

- Regulatory Guide 4.8, "Environmental Technical Specifications for Nuclear Power Plants," December 1975.

- Regulatory Guide 4.15, "Quality Assurance for Radiological Monitoring Programs (Normal Operations): Effluent Streams and the Environment," Rev. 1, February 1979.

U.S. Nuclear Regulatory Commission, "Radiological Assessment Branch Technical Position (BTP)," Rev. 1, Washington, DC, November 1979.

2.4   Criteria for Unrestricted Release — Non-Impacted Areas

For PSRs involving non-impacted areas, the preparation of the application is simpler, in that the licensee is not required to perform radiological surveys to identify and characterize the presence of residual radioactivity, nor to demonstrate compliance with 10 CFR 20.1402. The determination that an area is non-impacted is solely based on the results of an HSA and a current site characterization, taking into account facility operations and activities that may impact the area proposed for a PSR. Note that a licensee may voluntarily conduct a radiological survey to confirm the results of its HSA and include in the application a summary analysis of the results supporting the conclusion reported in the HSA.

The required information is described in 10 CFR 50.83(a)(1) and (a)(2), and 10 CFR 50.83(b)(1) – (b)(5). The requirements of 10 CFR 50.83 (a)(1) and (a)(2) are described in Section 2.3, "License Program Impacts." The requirements of 10 CFR 50.83(b)(1) – (b)(5) are described in Section 2.1, "General Information and Description of Area Proposed for Partial Site Release," Section 2.2, "Characterization of Area Proposed for Partial Site Release," and Section 2.7, "Environmental Review."

2.4.1 Areas of Review

The reviewer is responsible for confirming that the submitted information provides the means to determine that the area has not been impacted by prior site or facility operations, and is not likely to be impacted by future site or facility operations. The licensee's description of the area should be comprehensive to allow the staff to conclude, using the results of the HSA, a current characterization of the site, and MARSSIM guidance, that the area is correctly classified as non-impacted, as defined in 10 CFR 50.2.

2.4.2 Review Procedures

2.4.2.1 Acceptance Review

The NRC staff shall review and confirm that the application contains the information summarized under "Areas of Reviews," and described under "Information to be Submitted." The acceptance review shall confirm that the required information is complete to conduct a detailed technical analysis as the next step. The NRC staff shall confirm, via a preliminary review of the table of contents, content of each section, and any supporting attachments, that the licensee has provided the required information and determine that its level of detail is adequate. If the staff determines that the information is incomplete or insufficient to support a detailed technical analysis, the staff shall compile a list of such deficiencies and describe why the information is inadequate.

2.4.2.2 Safety Evaluation

The material provided by the licensee addresses specific technical information about the area(s) to be released. The staff shall review and evaluate the information at a level commensurate with the technical detail. The staff shall confirm that the licensee has used current site information, and that the characterization of the area proposed for PSR reflects its current conditions. The staff shall confirm that the licensee has provided the required information to allow independent confirmation of the conditions of the area to be released under 10 CFR 50.83(a) and (b), and that the staff has followed the review and evaluation process of 10 CFR 50.83(c) and (f) for non-impacted areas.

2.4.3 Acceptance Criteria

2.4.3.1 Regulatory Requirements

- 10 CFR 50.2, "Definitions."
- 10 CFR 50.83, "Release of Part of a Power Reactor Facility or Site for Unrestricted Use."

2.4.3.2 Regulatory Guidance

- NUREG-1575, Rev. 1, "Multi-Agency Radiation Survey and Site Investigation Manual (MARSSIM)":
  - Section 3.0, "Historical Site Assessment."
  - Section 4.4, "Classify Areas by Contamination Potential."

- NUREG-1757, Vol. 2, "Consolidated NMSS Decommissioning Guidance":
  - Section 1, "Purpose, Applicability, and Roadmap."
  - Appendix A.1, "Classification of Areas by Residual Radioactivity Levels."

2.4.3.3 Information To Be Submitted

For areas designated as non-impacted, the reviewer shall confirm that the licensee has completed an HSA of the area to be released, and the applicability of the information described in Section 2.2.3 (see "Acceptance Criteria") and information to be submitted for areas designated as radiologically non-impacted. In addition, the reviewer shall verify that the licensee has confirmed compliance with the following criteria:

- Doses to individual members of the public shall not exceed the limits and standards of Subpart D to 10 CFR Part 20.

- There is no reduction in the effectiveness of emergency planning or physical security.

- Effluent releases shall remain within the requirements of the conditions of the license.

- The REMP and ODCM are revised to account for changes associated with the partial site release.

- The site criteria of 10 CFR Part 100 will continue to be met.

- All other applicable statutory and regulatory requirements will continue to be met.

- The records associated with the PSR shall be maintained in accordance with 10 CFR 50.75(g).

- A schedule for the release of the property has been provided.

- The results of evaluations performed in accordance with 10 CFR 50.59 have been provided.

- A discussion provides the reasons for concluding that environmental impacts associated with the PSR are bounded by results issued in previously appropriate EISs.

If the licensee voluntarily conducted a radiological survey to confirm the results of its HSA, the reviewer should evaluate the supplemental information and assess whether it addresses the following considerations:

- description of survey methods used in assessing the radiological status of the area

- results of quality assurance/control measures used in conducting the associated radiation surveys, including sample collection and analysis

- discussion of the basis for assigned ambient background radioactivity and radiation levels in the area, or results reported during the conduct of surveys to confirm the radiological status of the area

- summary of radiological survey results and analysis of the results supporting the conclusion reported in the HSA

2.4.4 Evaluation Criteria and Findings

Reviews performed for this section of the SRP are based on guidance and criteria listed under "Regulatory Requirements" and "Regulatory Guidance," above. The reviewer shall verify that, where applicable, the licensee's application includes the information summarized under "Information to be Submitted," above. The staff's review shall verify, to the maximum extent practicable, that the information supplied by the licensee is complete and accurate by comparing it with prior licensee submissions, licensing actions, and inspection records maintained in NRC files. The reviewer shall verify that the characterization of the area is adequately described, and that the licensee's conclusions as to the non-impacted designation are supported by the HSA and a current site characterization, by appropriate resolutions of inconsistent information or data gaps in reconstructing the operational history of the site or facility, and that the process described in the referenced sections of NUREG-1575 and NUREG-1757 have been considered and used, as applicable. The reviewer shall confirm that the licensee meets the requirements of 10 CFR 50.83(a)(1) and (a)(2), and has provided the information required under 10 CFR 50.83(b)(1) – (b)(5).

2.4.5 References

U.S. Nuclear Regulatory Commission, NUREG-1575, "Multi-Agency Radiation Survey and Site Investigation Manual (MARSSIM)," Rev. 1, Washington, DC, August 2000.

U.S. Nuclear Regulatory Commission, NUREG-1757, "Consolidated NMSS Decommissioning Guidance, Characterization, Survey, and Determination of Radiological Criteria," Vol. 2, Final Report, Washington, DC, September 2003.

2.5   Criteria for Unrestricted Release — Impacted Areas

For PSRs involving impacted areas[4] and compliance with 10 CFR 20.1402, the required information is described in 10 CFR 50.83(a)(2) and (a)(3), and 10 CFR 50.83(d)(2). In meeting the criteria of 10 CFR 20.1402 for unrestricted release, one option is for the licensee to propose DCGLs corresponding to the annual dose limit of 0.25 mSv (25 mrem). DCGLs are derived for each of the radionuclides of concern expected to be present within the area considered for release. A number of options are available in selecting an approach, including the calculation of site-specific DCGLs, adopting the NRC's default screening DCGLs, or demonstrating compliance by calculating the dose that may be received by an average member of the critical group using the area once released based on the FSS results. The selection of a specific option should consider the intended use of the area after its release, features and conditions of the area to be released, potential impact of continued site operation on the area being released, whether other PSRs have occurred or are being contemplated before license termination, and considerations in planning the termination of the license for the entire site.

Because portions of the licensed site situated outside of the proposed area may not be addressed or remediated at the same time, there is a need to consider dose contributions from materials or activities from areas remaining in operation, from prior PSR areas on the derivation of an appropriate DCGL, or calculation of the dose based on the radiological conditions of the area once released. The DCGL should account for the movement of radioactivity under circumstances where the accumulation of contaminants could lead to increased radionuclide concentrations or would introduce exposure pathways different than considered initially. The movement of radioactivity may be associated with site activities (e.g., construction work, vehicular traffic, effluent releases, external radiation, etc.), and processes leading to movement of radioactivity (e.g., soil erosion, surface water drainage, or groundwater flow). Accordingly, the derivation of the DCGL should address all applicable transport mechanisms, given relevant site features, and include all associated exposure scenarios and pathways.

Another constraint that may impact the conditions of the PSR and continued operations of the facility is the dose limit associated with 10 CFR 20.1301(e). This requirement addresses compliance with 40 CFR Part 190, which limits the total dose that a member of the public may receive from all fuel cycle facilities. In showing compliance with 10 CFR 20.1301(e) for the remaining portions of the site, the dose from the PSR and any radiological effluents and external radiation from site operations must be combined to show compliance. As opposed to the requirements of 10 CFR 20.1401 and 20.1402, this constraint is based on doses to an actual receptor. Accordingly, a licensee should consider the requirements of both criteria in defining the most appropriate exposure scenarios and associated pathways in deriving the associated DCGLs.

---

[4]The definition of "impacted areas" can be found in 10 CFR 50.2, with additional details provided in NUREG-1575 and NUREG-1757.

In the context of 10 CFR 20.1402, the dose modeling for a PSR may be grouped into the following five general steps:

(1)   source term development

(2)   definition and selection of critical groups and exposure scenarios and pathways

(3)   formulation of the conceptual dose models

(4)   sensitivity and uncertainty analysis

(5)   process used for demonstrating compliance with criteria

The option of using the NRC's default screening DCGLs may not be available in some instances because of sit-specific conditions. Conditions that would exclude the use of screening DCGLs include contaminated soils with radioactivity at depth greater than 30 cm (12 in.), radioactivity present in surface and groundwater, building with volumetrically contaminated materials, radioactivity present in surface water sediments, and sites with geohydrological conditions resulting in surface water runoff.

In light of these complexities, this section outlines prerequisites and refers to the more exhaustive guidance presented in NUREG-1757 and other supporting documents, rather than repeating it in its entirety. The documents listed under "Regulatory Guidance," below, identify the sections of NUREG-1757 where specific information may be found. Finally, the NRC encourages licensees to discuss the approach being considered in planning the PSR, as well as the basis and assumptions used modeling DCGLs, with the staff before finalizing the application package.

2.5.1   Areas of Review

As part of its review, the staff shall evaluate the basis of the radiological source term; definition and selection of critical groups, and exposure scenarios and pathways; rationale for using the NRC's default screening DCGLs; basis of the conceptual dose models in calculating site-specific DCGLs; and results of the uncertainty analysis. The reviewer shall determine whether the dose model description and supporting information adequately demonstrate compliance with the radiological criteria for unrestricted release. For certain cases, such as screening analyses using default DCGLs, the licensee is expected to submit sufficient information to demonstrate that site features, model parameters, and exposure scenarios and pathways are generally consistent with those forming the basis of screening DCGLs. In addition, the reviewer shall evaluate the ALARA analysis, which is based, in part, on dose models and model assumptions. Finally, the reviewer shall evaluate the information documenting the process used to comply with 10 CFR 20.1301(e) for fuel cycle facilities.

2.5.2   Review Procedures

2.5.2.1 Acceptance Review

The NRC staff shall review and confirm that the application contains the information summarized under "Areas of Reviews," and described under "Information to be Submitted." The acceptance review shall confirm that the required information is complete to conduct a detailed technical analysis as the next step. The NRC staff shall confirm, via a preliminary review of the table of contents, content of each section, and any supporting attachments, that the licensee has provided the required information and determine that its level of detail is adequate. If the staff determines that the information is incomplete or insufficient to support a detailed technical analysis, the staff shall compile a list of such deficiencies and describe why the information is inadequate.

2.5.2.2 Safety Evaluation

The material provided by the licensee addresses specific technical and regulatory topics in developing site-specific unrestricted release criteria or using the NRC's default release criteria. The staff shall review and evaluate the information at a level commensurate with the technical detail. The staff shall confirm that the licensee has used defensible information, and that descriptions of the approach, models, assumptions, and results are valid and can withstand scrutiny against the technical and regulatory requirements of 10 CFR 50.83(a)(3) and (d)(2), and 10 CFR 20.1402. The staff shall confirm that the licensee has provided the required information to allow independent confirmation of the licensee's proposed unrestricted release criteria and conclusions that the PSR will meet the requirements of 10 CFR 50.83(a)(3) and (d)(2) for impacted areas, and that the staff has followed the review and evaluation process of 10 CFR 50.83(e) and (f) for impacted areas.

2.5.3   Acceptance Criteria

2.5.3.1 Regulatory Requirements

- 10 CFR 20.1301(e), "Dose Limits for Individual Members of the Public."
- 10 CFR 20.1402, "Radiological Criteria for Unrestricted Use."

2.5.3.2 Regulatory Guidance

- NUREG-1757, Vol. 2, "Consolidated NMSS Decommissioning Guidance":
  - Section 2.5, "Demonstrating Compliance Using Dose Assessment Methods Versus Derived Concentration Guideline Levels and Final Status Surveys."
  - Section 3.4, "Considerations for Other Constraints on Allowable Residual Radioactivity."
  - Section 5.0, "Dose Modeling Evaluation."
  - Section 6.0, "ALARA Analysis."
  - Appendix H, "Criteria for Conducting Screening Dose Modeling Evaluations."
  - Appendix I, "Technical basis for Site Specific Dose Modeling Evaluations."
  - Appendix K, "Dose Modeling Considerations for Partial Site Release."
  - Appendix L, "Worksheet for Identifying Potential Pathways for Partial Site Release."
  - Appendix N, "ALARA Analyses."

- NUREG/CR-5512, Vol. 1, "Residual Radioactive Contamination from Decommissioning, Technical Basis for Translating Contamination Levels to Annual Total Effective Dose Equivalent." **Note:** See Vol. 3 of NUREG/CR-5512 for parameter values deemed acceptable by the NRC staff.

- NUREG/CR-5512, Vol. 2, "Residual Radioactive Contamination from Decommissioning, User's Manual, DandD Version 2.1."

- NUREG/CR-5512, Vol. 3, "Residual Radioactive Contamination from Decommissioning, Parameter Analysis."

- NUREG/CR-6676, "Probabilistic Dose Analysis Using Parameter Distributions Developed for RESRAD and RESRAD-BUILD Codes."

- NUREG/CR-6692, "Probabilistic Modules for the RESRAD and RESRAD-BUILD Computer Codes."

- NUREG/CR-6697, "Development of Probabilistic RESRAD 6.0 and RESRAD-BUILD 3.0 Computer Codes."

- ANL/EAD-4, "User's Manual for RESRAD Version 6."

- Federal Guidance Report No. 11, "Limiting Values of Radionuclide Intake and Air Concentration and Dose Conversion Factors for Inhalation, Submersion, and Ingestion."

- Federal Guidance Report No. 12, "External Exposure to Radionuclides in Air, Water, and Soil."

2.5.3.3 Information To Be Submitted

In considering dose modeling, either for screening or site-specific release criteria for site grounds, the reviewer shall verify that the licensee has provided the following information:

- discussion of how the dose modeling option addresses potential effects attributable to current and future site operations or previous PSRs

- discussion of whether the PSR may impact the license termination of the licensed site, including the implementation of any additional PSRs

- discussion of potential sources of radiation exposures from licensed activities and whether they have been constrained or remediated[5] in light of interactions with the remaining portions of the licensed site or previous PSRs, including results of ALARA analyses

- if screening criteria are used, an analysis showing how they have been adjusted to account for any considerations associated with the proposed PSR

- documentation demonstrating compliance with the requirements of Section 5.1 of NUREG-1757 (Vol. 2) when applying the NRC's default screening criteria

- documentation demonstrating compliance with the requirements of Section 5.2 of NUREG-1757 (Vol. 2) when deriving site-specific criteria

---

[5]See NUREG-1757, Vol. 2, App. K, for specific details on dose modeling approaches and considerations.

In considering the use of the NRC's default screening criteria for buildings, the reviewer shall verify that the licensee has provided the following information:

- the general conceptual model for both the source term and building environment

- a summary of the screening method used (i.e., running DandD or using lookup tables in App. H to NUREG-1757, Vol. 2)

In considering the use of the NRC's default screening criteria for surface soils, the reviewer shall verify that the licensee has provided the following information:

- justification on the appropriateness of using the screening approach for both the source term and the environment

- confirmation that the site is not characterized with conditions that would exclude the use of a screening approach, such as a site with unique conditions

- a summary of the screening method used (i.e., running DandD or using look-up tables in App. H to NUREG-1757, Vol. 2)

In considering the derivation and application of site-specific release criteria for buildings or soils (surface and subsurface), the reviewer shall verify that the licensee has provided the following information:

- source term information including radionuclides of interest, configuration of the source, variability of the source, etc.

- description of exposure scenarios, including a description of the critical group

- description of the conceptual model of the site, including the source term, and physical features important in modeling transport pathways for members of the critical group

- identification and description of the mathematical model used [e.g., hand calculations, DandD Screen Ver. 2.1, RESRAD (Ver. 6.3), RESRAD-BUILD (Ver. 3.3), or other analytical method]

- description of the parameters used in the analysis

- discussion of the effect of uncertainty on the results

- input and output files or printouts, if computer programs were used

In considering ALARA, the reviewer shall verify that the licensee has provided the following information:

- description of how ALARA practices were applied in meeting the DCGLs

- basis for concluding that the dose to the average member of the critical group is ALARA

- description of methods used to remediate or mitigate residual radioactivity levels below DCGLs through good radiological work practices, such as routine housekeeping and maintenance

- results of cost-benefit analyses, if necessary, supporting the conclusion that doses are ALARA

- description of model assumptions and parameters that exemplify the use of realistic conservatism in forming the basis of generic screening or site-specific DCGLs

2.5.4 Evaluation Criteria and Findings

Reviews performed for this section of the SRP are based on guidance and criteria listed under "Regulatory Requirements" and "Regulatory Guidance," above. The reviewer shall verify that, where applicable, the licensee's application includes the information summarized under "Information to be Submitted," above. The staff's review shall verify, to the maximum extent practicable, that the information supplied by the licensee is complete and demonstrates compliance with 10 CFR 20.1402 and 10 CFR 20.1301(e) for facilities that are part of the fuel cycle under the EPA's environmental radiation standards (40 CFR Part 190). The reviewer shall verify that the information is provided and presented in a manner consistent with the above-referenced sections of NUREG-1757 and supporting documents. Appendices I, K, and L to NUREG-1757 (Vol. 2) present detailed guidance in structuring such reviews.

Regarding ALARA, the reviewer shall verify that the information and supporting analyses provide reasonable assurance that future activities taking place on the released site and remediation activities conducted for the PSRwill result in doses that are ALARA. For example, ALARA practices include, among others, the use of good housekeeping and maintenance practices, as described in NUREG-1757 (Vol. 2, Section 6). In light of the conservatism incorporated in the derivation of the NRC's generic screening DCGLs for building surfaces and surface soils, the reviewer may presume, absent information to the contrary, that licensees that have remediated building surfaces or soils to generic screening DCGLs do not need to demonstrate that these levels are ALARA. Moreover, the "Statements of Consideration" for Subpart E of 10 CFR Part 20 and the Final Generic Impact Statement (NUREG-1496, NRC 1997) show that the removal of soils for offsite disposal is not cost-effective [i.e., resulting in an annual dose lower than 25 mrem (0.25 mSv)] for unrestricted release exposure scenarios.

2.5.5 References

U.S. Environmental Agency, "Limiting Values of Radionuclide Intake and Air Concentration and Dose Conversion Factors for Inhalation, Submersion, and Ingestion," Federal Guidance Report No. 11, EPA 520/1-88-020, Washington, DC, September 1988.

U.S. Environmental Agency, "External Exposure to Radionuclides in Air, Water, and Soil," Federal Guidance Report No. 12, EPA 402-R-93-081, Washington, DC, September 1993.

U.S. Nuclear Regulatory Commission, NUREG/CR-5512, "Residual Radioactive Contamination from Decommissioning, Technical Basis for Translating Contamination Levels to Annual Total Effective Dose Equivalent," Vol. 1, Final Report, Washington, DC, October 1992.

U.S. Nuclear Regulatory Commission, NUREG/CR-5512, "Residual Radioactive Contamination from Decommissioning, User's Manual, DandD Version 2.1," Vol. 2, Washington, DC, April 2001.

U.S. Nuclear Regulatory Commission, NUREG/CR-5512, "Residual Radioactive Contamination from Decommissioning, Parameter Analysis," Vol. 3, Draft, Washington, DC, October 1999.

U.S. Nuclear Regulatory Commission, NUREG/CR-6676, "Probabilistic Dose Analysis Using Parameter Distributions Developed for RESRAD and RESRAD-BUILD Codes," Washington, DC, July 2000.

U.S. Nuclear Regulatory Commission, NUREG/CR-6692, "Probabilistic Modules for the RESRAD and RESRAD-BUILD Computer Codes," Washington, DC, November 2000.

U.S. Nuclear Regulatory Commission, NUREG/CR-6697, "Development of Probabilistic RESRAD 6.0 and RESRAD-BUILD 3.0 Computer Codes," Washington, DC, December 2000.

U.S. Nuclear Regulatory Commission, NUREG-1575, "Multi-Agency Radiation Survey and Site Investigation Manual (MARSSIM)," Rev. 1, Washington, DC, August 2000.

U.S. Nuclear Regulatory Commission,NUREG-1757, "Consolidated NMSS Decommissioning Guidance, Characterization, Survey, and Determination of Radiological Criteria," Vol. 2, Final Report, Washington, DC, September 2003.

U.S. Department of Energy, "User's Manual for RESRAD Version 6," ANL/EAD-4, Argonne National Laboratory, Environmental Assessment Division, Argonne, Illinois, July 2001.

2.6   Final Status Survey Design and Final Status Survey Results

For PSRs involving impacted areas and compliance with 10 CFR 20.1402, the required information is described in 10 CFR 50.83(a)(3) and (d)(2). The licensee is required to conduct radiological surveys to demonstrate compliance with the radiological criteria for unrestricted use of 10 CFR 20.1402. The area being released may be divided into discrete survey units to facilitate the FSSs. The size of each survey unit, in part, depends on the total surface area of the portion being released, survey classification, and other features (e.g., physical and radiological). FSSs are performed to demonstrate that residual radioactivity levels in survey units meet the release criteria corresponding to the annual dose limit of 0.25 mSv (25 mrem). The release criteria (DCGLs) are expressed either as pCi/g (Bq/g) for activity distributed volumetrically, or dpm/100 cm^2 (Bq/100 cm^2) for activity distributed on surfaces of remaining structures and buildings. The development of FSSs is based on prior information, such as from HSAs, process knowledge, and results of characterization or post-remediation surveys. An important aspect of the design process involves the use of data quality objectives (DQOs), which identify constraints and criteria for statistical tests used in deciding whether the survey unit meets release criteria. The MARSSIM and NUREG-1757 (Vol. 2) present methods acceptable to the NRC for conducting surveys in areas where contaminants are present in soils (surface and subsurface) and on surfaces of structures and buildings. However, the NRC recognizes that alternative methods may be used to achieve the same objective.

Given the complexity of the guidance presented in MARSSIM, this section outlines its major elements and refers to the more exhaustive guidance presented in cited references, rather than replicating it here. The documents listed under "Regulatory Guidance," identify sections of NUREG-1757, NUREG-1575, NUREG-1576, NUREG-1505, and NUREG-1507 where specific information may be found. Finally, the NRC encourages licensees to discuss with staff the approach being considered in designing and implementing FSSs using MARSSIM or an alternative method.

2.6.1 Areas of Review

Final status surveys are performed in areas that have been designated as impacted and are classified according to the potential for residual radioactivity. Under MARSSIM, impacted areas may be classified as Class 1, 2, or 3 areas. Class 1 areas necessitate more stringent survey requirements, while those for Class 2 and 3 areas are progressively less demanding. The criteria defining impacted areas and survey unit classifications are presented in NUREG-1757 and NUREG-1575. For subsurface soils, additional guidance, beyond that of MARSSIM, may be found in NUREG-1757. FSSs are conducted in areas that have been fully characterized, remediated if necessary, and found to meet all elements of the DQO process. Although the FSS is discussed as if it were an activity performed at a single stage in the process of releasing an area, this is not always the case. Data from other surveys, such as characterization and remedial action support surveys, can provide valuable information in designing FSSs, provided they meet specific aspects of the DQO process. FSSs are designed and implemented using the MARSSIM-based classification system, including survey grid basis, surface scan coverage, sampling locations, number of sampling points, number and depth of soil core samples, and conduct of direct survey measurements. The FSS process provides data to demonstrate that survey results satisfy the DCGLs for the radionuclides of concern, survey conditions, and DQOs. In documenting survey results, the final status survey report (FSSR) should stand on its own with minimal information incorporated by reference.

For non-impacted areas, the conduct of FSS is not mandated by MARSSIM and NRC guidance. However, licensees have the option to provide survey results tp supplement the HSA results in order to confirm that the area slated for PSR is indeed not impacted. The design and conduct of such surveys need not be based on MARSSIM, as licensees may use alternative methods. Given that this information would be submitted voluntarily, the PM and reviewers may identify only particular elements of the survey design and results, and define the corresponding level of technical review. In such instances, the staff may not carry out all review steps given below.

2.6.2 Review Procedures

2.6.2.1 Acceptance Review

The NRC staff shall review and confirm that the application contains the information summarized under "Areas of Reviews," and described under "Information to be Submitted." The acceptance review shall confirm that the required information is complete to conduct a detailed technical analysis as the next step. The NRC staff shall confirm, via a preliminary review of the table of contents, content of each section, and any supporting attachments, that the licensee has provided the required information and determine that its level of detail is adequate. If the staff determines that the information is incomplete or insufficient to support a detailed technical analysis, the staff shall compile a list of such deficiencies and describe why the information is inadequate.

2.6.2.2 Safety Evaluation

The material provided by the licensee addresses specific technical and regulatory topics to demonstrate that the proposed site-specific unrestricted use criteria or default NRC screening criteria have been met, given the results of final status radiological surveys of the area(s) to be released. The staff shall review and evaluate the information at a level commensurate with the technical detail. The staff shall confirm that the licensee has used defensible information, and that descriptions supporting the design, planning, execution, and evaluations of FSS results meet the technical and regulatory requirements of 10 CFR 50.83. The staff shall confirm that the licensee has provided the required information and FSS results to allow independent confirmation that the licensee's conclusions are valid and the evaluation demonstrates compliance with the requirements of 10 CFR 50.83(a)(3) and (d)(2) for impacted areas, and that the staff has followed the review and evaluation process of 10 CFR 50.83(e) and (f) for impacted areas.

2.6.3 Acceptance Criteria

2.6.3.1 Regulatory Requirements

- 10 CFR 20.1402, "Radiological Criteria for Unrestricted Use."

- 10 CFR 50.83, "Release of Part of a Power Reactor Facility or Site for Unrestricted Use."

2.6.3.2 Regulatory Guidance

- NUREG-1575, "Multi-Agency Radiation Survey and Site Investigation Manual (MARSSIM)":
    - Section 4, "Preliminary Survey Considerations."
    - Section 5, "Survey Planning and Design."
    - Section 6, "Field Measurement Methods and Instrumentation."
    - Section 7, "Sampling and Preparation for Laboratory Measurements."
    - Section 8, "Interpretation of Survey Results."
    - Section 9, "Quality Assurance and Control."
    - Appendix D, "The Planning Phase of the Data Life Cycle."
    - Appendix E, "The Assessment Phase of the Data Life Cycle."
    - Appendix I, "Statistical Tables and Procedures."

- NUREG-1757, Vol. 2, "Consolidated NMSS Decommissioning Guidance":
    - Section 4.4, "Final Status Survey Design."
    - Section 4.5, "Final Status Survey Report."
    - Appendix A, "Implementing the MARSSIM Approach for Conducting Final Radiological Surveys."
    - Appendix D, "Survey Data Quality and Reporting."
    - Appendix E, "Measurements for Facility Radiation Surveys."
    - Appendix G.2.1, "Subsurface Residual Radioactivity."

- NUREG-1507, "Minimum Detectable Concentrations with Typical Radiation Survey Instruments for Various Contaminants and Field Conditions":
  - Section 3, "Statistical Interpretations of Minimum Detectable Concentrations."
  - Section 4, "Variables Affecting Instrument Minimum Detectable Concentrations."
  - Section 5, "Variables Affecting Minimum Detectable Concentrations in the Field."
  - Section 6, "Human Performance and Scanning Sensitivity."

- NUREG-1576, Vol. 2, "Multi-Agency Radiological Laboratory Analytical Protocols Manual (MARLAP)":
  - Section 10, "Field Sampling Issues That Affect Laboratory Measurements."

- NUREG-1505, "A Nonparametric Statistical Methodology for the Design and Analysis of Final Status Decommissioning Surveys."

2.6.3.3 Information To Be Submitted

In considering the design of FSS, the reviewer shall verify that the licensee has provided the following information:

- overview describing the FSS design and associated DQOs

- description and map(s) or drawing(s) of impacted areas of the site, area, or building classified by residual radioactivity levels (Class 1, Class 2, or Class 3) and divided into survey units, and basis for their configurations into survey units (maps should have compass headings and be referenced to a geodetic site plan survey marker)

- description of the background reference areas and materials, if used, and justification for their selection

- basis of the $DCGL_W$ applied to the area slated for PSR and the application of any constraints on DCGL values

- for surface soils, a summary of the statistical tests [Sign or Wilcoxon Rank Sum (WRS) test] that will be used to evaluate survey results, including the elevated measurement comparison test in Class 1 survey units, a justification for any test methods not included in MARSSIM, and probabilities for decision errors (α for Type I and β for Type II), including a justification for an α probability greater than 0.05

- for subsurface soils, a justification of the method used in defining the number and depths of soil core samples, a summary of the statistical tests (Sign or WRS test) used to evaluate survey results, basis of core sampling scheme and homogenization over depths consistent with dose assessment models, criteria defining the elevated measurement comparison test over the specified core depth in Class 1 survey units, and probabilities for decision errors (α for Type I and β for Type II), including a justification for an α probability greater than 0.05

- for surfaces associated with structures and buildings, a summary of the statistical tests (Sign or WRS test) that will be used to evaluate survey results, including the elevated measurement comparison test in Class 1 survey units, a justification for any test methods not included in MARSSIM, and probabilities for decision errors (α for Type I and β for Type II), including a justification for an α probability greater than 0.05

- description of scanning instruments, methods, calibration, operational checks, scan coverage, and scanning minimum detectable concentration (MDC) for each media and radionuclides or surrogate radionuclides

- description of the instruments, calibration, operational checks, and MDC used for *in situ* direct measurements, with a demonstration that instruments and methods have the proper sensitivity in detecting activity levels at the $DCGL_W$ and investigational levels

- description of the basis defining sampling locations, sampling pattern, and sample size within each survey unit, including the grid reference system and random start systematic sample locations for Class 1 and 2 survey units, and random locations shown for Class 3 survey units and reference areas

- description of the analytical instruments for measuring samples in the laboratory, including the calibration, minimum detectable activity (MDA), and methodology for evaluation, with a demonstration that instruments and survey methods have the proper sensitivity

- description of the quality assurance (QA) project plan commensurate with DQOs for survey measurements and laboratory sample analysis, including data quality indicators (DQIs), such as precision, accuracy (bias), representativeness, comparability, and completeness (PARCC)

- description of how samples to be analyzed in the laboratory will be collected, controlled, and handled

- description of the FSS investigation levels and how they were determined

In considering the FSSR, the reviewer shall verify that the licensee has provided the following information:

- overview of the results of the FSS conducted over site grounds (surface and subsurface soils) and surfaces of structures and buildings

- summary of the basis of the DCGLs applied to the area slated for PSR and the application of any constraints on DCGL values

- description of the method by which the number of samples, sampling locations, and sampling patterns were determined for each survey unit in the area

- survey results for each survey unit, including the following:

  - number of surface soil samples and/or subsurface soil core samples or direct measurements taken in each survey unit

  - number of direct surface measurements or samples taken in each survey unit on surfaces of structures and buildings

  - description of the survey unit, including (1) a map or drawing of the survey unit showing the site reference system and random start systematic sample locations for Class 1 and 2 survey units, and random locations for Class 3 survey units and reference areas, (2) discussion of remedial actions and unique features of the survey unit, and (3) delineation of areas scanned and scan coverage for Class 2 and 3 survey units

- sample concentration or measurement results expressed as final radiological units comparable to the DCGL

- results of statistical evaluations of sample concentrations or direct measurement results using the Sign or WRS test

- results of judgmental and miscellaneous survey results reported separately from those measurements and samples collected and used in MARSSIM statistical evaluations

- discussion of anomalous results, including areas with elevated direct radiation levels noted during scanning in excess of investigation levels or the $DCGL_W$

- for any sample point exceeding the $DCGL_W$, a statement confirming that the survey unit satisfied the $DCGL_W$ and elevated measurement comparison test

- results of QA and quality control (QC) evaluations associated with the validation of FSS results and laboratory sample analysis results

- description of any changes in initial survey unit design assumptions (e.g., classification, size, etc.) relative to the extent and distribution of residual radioactivity

2.6.4 Evaluation Criteria and Findings

Reviews performed for this section of the SRP are based on guidance and criteria listed under "Regulatory Requirements" and "Regulatory Guidance," above. The reviewer shall verify that, where applicable, the licensee's application includes the information summarized under "Information to be Submitted," above. The staff's review shall verify, to the maximum extent practicable, that the information supplied by the licensee is complete and sufficient to demonstrate compliance with 10 CFR 20.1402 and 10 CFR 50.83(a)(3) and (d)(2).

For results indicating that a survey unit had failed initially, was subsequently remediated, and was then shown to meet the $DCGL_W$ after the conduct of another FSS, the reviewer shall assess (1) whether the information properly describes the investigation that was conducted to ascertain the reason for the failure, and (2) evaluate the impact of that failure on the initial conclusion that the area was characterized properly and ready for the conduct of FSSs.

The reviewer shall verify that the information is provided and presented in a manner consistent with the above referenced sections of NUREG-1757, NUREG-1575, NUREG-1507, NUREG-1505, and NUREG-1576. The review shall verify that the FSS design demonstrates compliance with the radiological release criteria for surface and subsurface soils, and surfaces of any remaining structures and buildings. The FSS design meets the evaluation criteria in NUREG-1757, Vol. 2, Section 4.4. The review shall verify that the licensee's results presented in the FSSR support the conclusion that the survey unit(s) comprised within the area slated for PSR meet(s) the radiological criteria of 10 CFR 20.1402. The FSSR results are adequate if they meet the evaluation criteria of NUREG-1757, Vol. 2, Section 4.5.

2.6.5 References

U.S. Nuclear Regulatory Commission, NUREG-1496, "Generic Environmental Impact Statement in Support of Rulemaking on Radiological Criteria for License Termination of NRC-Licensed Nuclear Facilities," Final Report, Washington, DC, July 1997.

U.S. Nuclear Regulatory Commission, NUREG-1505, "A Nonparametric Statistical Methodology for the Design and Analysis and Final Status Decommissioning Surveys," Rev. 1, Draft Report, Washington, DC, June 1998.

U.S. Nuclear Regulatory Commission, NUREG-1507, "Minimum Detectable Concentrations with Typical Radiation Survey Instruments for Various Contaminants and Field Conditions," Washington, DC, June 1998.

U.S. Nuclear Regulatory Commission, NUREG-1575, "Multi-Agency Radiation Survey and Site Investigation Manual (MARSSIM)," Rev. 1, Washington, DC, August 2000.

U.S. Nuclear Regulatory Commission, NUREG-1576, "Multi-Agency Radiological Laboratory Analytical Protocols Manual (MARLAP)," Washington, DC, July 2004.

U.S. Nuclear Regulatory Commission, NUREG-1757, "Consolidated NMSS Decommissioning Guidance, Characterization, Survey, and Determination of Radiological Criteria," Vol. 2, Final Report, Washington, DC, September 2003.

2.7 Environmental Review

A PSR may involve changes in the configuration of the site boundary and facility operations. In turn, such changes may result in environmental impacts. The types of impacts can vary, depending on the size of the parcel of land being released and how that parcel of land was affected by prior and current facility operations. For example, changes to the site boundary or portions of the site defined by the license might affect environmental resources and current and future land uses in the vicinity of the site. The following requirements address environmental reviews (ERs) and consider both impacted and non-impacted areas:

- For impacted areas, 10 CFR 50.83(d)(3) requires the licensee to submit a supplement to the environmental report, under 10 CFR 51.53, describing any new information or significant environmental changes associated with the proposed release of the property.

- For non-impacted areas, 10 CFR 50.83(b)(5) requires the licensee to submit a discussion that provides reasons for concluding that environmental impacts associated with the proposed release of the property is bounded by a previously issued appropriate EIS.

Partial site releases are analogous to license terminations, with respect to the portion of the site to be released. Accordingly, the provisions of 10 CFR 51.53(d), describing any new information or significant environmental change associated with the licensee's proposed license termination activities, should also be applied to PSRs. In more specific terms, 10 CFR 51.53(d) addresses the post-operating license stage of a facility, given that its related impacts are similar to those associated with license termination.

The licensee should conduct an ER that evaluates and characterizes potential impacts, including how changes to the site boundary or licensed site might affect environmental resources and current and future land uses around the site. The ER should include a detailed evaluation of each affected resource in assessing potential impacts. In assessing impacts, it may not be necessary to evaluate each resource to the same level of detail, and some resources may not require any specific evaluation or reevaluation. For example, a PSRis not expected to impact site meteorology, climatology, seismology, geology, and geotechnical characteristics of the site, and, therefore, the application would not normally address these resources. By contrast, a review of license conditions or radiological effluent technical specifications may identify specific aspects that may be impacted by the PSR and would need to be addressed in the evaluation for either an impacted or non-impacted area.

Finally, the licensee should assess whether the proposed PSR might impact ongoing activities of the REMP, and, if so, identify the required changes to the program while conforming with Appendix I to 10 CFR Part 50 and the ODCM. The licensee shall confirm that changes in the description of the facility and operations that could result in significant environmental changes were evaluated and addressed in compliance with license conditions and FSAR or USAR. The specific requirements are addressed in Section 2.3, "License Program Impacts," and are not repeated here.

2.7.1 Areas of Review

The PSR regulations require licensees to address potential impacts on the environment and local resources as a result of releasing part of the property and associated changes to current site boundaries. For a non-impacted area, the licensee must state the reasons for concluding that the environmental impacts associated with the proposed release will be bounded by an appropriate previously issued EIS.

For an impacted area, the request shall be submitted in the form of a license amendment application. The licensee is required to provide a supplement to the existing environmental report, pursuant to 10 CFR 51.53(d), describing any new information or significant environmental changes associated with the licensee's proposed PSR. The PM and technical reviewers shall contact the Environmental Review Section in NMSS/DWMEP in addressing the requirements of 10 CFR 51.53(d) as they apply to a specific reactor site and in evaluating the information submitted by the licensee.

If the proposed PSR is expected to impact ongoing activities of the REMP, the licensee shall address changes to the program in accordance with Appendix I to 10 CFR Part 50; Regulatory Guides 4.1, 4.8, and 4.15; and the Radiological Assessment BTP, taking into account site-specific characteristics. See Section 2.3, "License Program Impacts," for further details. Any changes shall be documented and incorporated in a revised ODCM.

If groundwater or soil is known or suspected to be contaminated with plant-derived radioactivity, the PM shall consider the provisions of the MOU between the EPA and NRC in identifying their respective roles for decommissioning NRC-licensed sites (*Federal Register*, Vol. 67, No. 206, p.65375, Oct. 24, 2002). The MOU includes provisions for joint consultations for sites when at the time of license termination (1) groundwater contamination exceeds the EPA's allowable levels, (2) the NRC contemplates restricted release or alternative criteria for the site, or (3) residual radioactive soil concentrations exceed the levels defined in the MOU. The PM and technical reviewers shall contact the Reactor Decommissioning Section in NMSS/DWMEP in addressing the implementation of the MOU at a specific reactor site. In addition, NUREG-1757 provides supporting guidance that may be used in evaluating such sites (see Vol. 1, Section 9.3, and Vol. 2, Appendices F, G.2, H, and K).

The licensee shall also comply with environmental and health protection regulations governing the presence of any other toxic or hazardous properties of materials present on the area proposed for PSR. The presence of toxic or hazardous materials may be associated with spills and leaks involving the use of industrial chemicals, reagents, lubricants, etc. The licensee should describe the types of permits and authorizations needed from other Federal and State and local regulatory agencies, as required.

2.7.2   Review Procedures

2.7.2.1 Acceptance Review

The NRC staff shall review and confirm that the application contains the information summarized under "Areas of Reviews," and described under "Information to be Submitted." The acceptance review shall confirm that the required information is complete to conduct a detailed technical analysis as the next step. The NRC staff shall confirm, via a preliminary review of the table of contents, content of each section, and any supporting attachments, that the licensee has provided the required information and determine that its level of detail is adequate. If the staff determines that the information is incomplete or insufficient to support a detailed technical analysis, the staff shall compile a list of such deficiencies and describe why the information is inadequate.

2.7.2.2 Safety Evaluation

The material provided by the licensee addresses specific technical and regulatory topics describing potential impacts of a PSR on the site and its environs. The staff shall review and evaluate the information at a level commensurate with the technical detail. The staff shall confirm that the licensee has used defensible information, and descriptions of the site and its environs are currently valid. The staff shall confirm that the licensee has provided the required information to allow independent confirmation of the licensee's assessment of the types and extent of impacts. The information describing potential impacts should address the following considerations:

- land use
- population demographics
- socioeconomic factors
- environmental justice

- visual or scenic factors

- transportation

- air quality

- noise

- public and occupational health

- surface and groundwater

- natural resources

- ecology

If groundwater or soil is known or suspected to be contaminated with plant-derived radioactivity, the PM shall consider the provisions of the MOU between the EPA and NRC in identifying their respective roles for decommissioning NRC-licensed sites (*Federal Register*, Vol. 67, No. 206, p. 65375, Oct. 24, 2002). The PM and technical reviewers shall contact the Reactor Decommissioning Section in NMSS/DWMEP in addressing the implementation of the MOU at a specific reactor site. A determination shall be made as to whether (1) the portion of the site designated for release is eligible for a PSR in its current or expected radiological condition given surface and subsurface soil transport mechanisms, (2) remediation might be a feasible option in mitigating the presence of contaminated soil and groundwater, and (3) the PSR action might be deferred to the time of license termination and implemented as a staged release under an approved LTP.

The review shall confirm that the stated impacts are bounded by prior EAs and meet the requirements of 10 CFR 50.83(a) and (b)(5) for non-impacted areas, or 10 CFR 50.83(a) and (d)(3) for impacted areas, and that the staff has followed the review and evaluation process of 10 CFR 50.83(c) and (f) for non-impacted areas, or 10 CFR 50.83(e) and (f) for impacted areas.

2.7.3 Acceptance Criteria

2.7.3.1 Regulatory Requirements

- 10 CFR Part 20, "Standards for Protection Against Radiation."

- 10 CFR 51.53, "Post-Construction Environmental Reports."

- 10 CFR 51.53(d), "Post-Operating License Stage."

- 10 CFR Part 50, "Domestic Licensing of Production and Utilization Facilities."

- 10 CFR 50.82, "Termination of License."

- 10 CFR 50.83, "Release of Part of a Power Reactor Facility or Site for Unrestricted Use."

- 10 CFR Part 50, "Domestic Licensing of Production and Utilization Facilities," Appendix I, "Numerical Guides for Design Objectives and Limiting Conditions for Operation to Meet the Criterion 'As Low As Is Reasonably Achievable' for Radioactive Material in Light-Water-Cooled Nuclear Power Reactor Effluents."

- 10 CFR 51.55, "Environmental Report — Distribution."

- 10 CFR Part 72, "Licensing Requirements for the Independent Storage of Spent Nuclear Fuel, High-Level Radioactive Waste, and Reactor-Related Greater Than Class C Waste."

- 10 CFR Part 100, "Reactor Site Criteria."

2.7.3.2 Regulatory Guidance

- NRR Office Instruction LIC-203, "Procedural Guidance for Preparing Environmental Assessments and Considering Environmental Issues."

- NUREG-1748, "Environmental Review Guidance for Licensing Actions Associated with NMSS Programs."

- NUREG-1555, "Standard Review Plans for Environmental Reviews for Nuclear Power Plants."

- Regulatory Guide 4.1, "Programs for Monitoring Radioactivity in the Environs of Nuclear Power Plants."

- Regulatory Guide 4.8, "Environmental Technical Specifications for Nuclear Power Plants."

- Regulatory Guide 4.15, "Quality Assurance for Radiological Monitoring Programs (Normal Operations): Effluent Streams and the Environment."

- U.S. Nuclear Regulatory Commission, "Radiological Assessment Branch Technical Position (BTP)."

2.7.3.3 Information To Be Submitted

For a PSR involving a radiologically impacted area, the licensee shall submit a license amendment request and a supplement to the existing environmental report, pursuant to 10 CFR 50.83(d)(3) and 10 CFR 51.53(d), describing any new information or significant environmental changes associated with the proposed release of the property. The reviewer shall verify that the licensee has provided the following information:

- description of potential impacts on the environment, local resources, and changes to the site boundary, including site and area maps outlining impacts by types and locations

- description of unique site- or facility-specific issues and information not identified in prior reports submitted to the NRC

- delineation of the area designated for PSR and the radiological basis for its impacted status

- compliance with environmental and health protection regulations governing the presence of any other toxic or hazardous properties of materials present on the area proposed for PSR

- proposed schedule for the PSR

- analysis of information and rationale leading to conclusions about the nature and scope of environmental impacts associated with the proposed release

For a PSR involving a non-impacted area, the licensee shall provide, pursuant to 10 CFR 50.83(b)(5), a discussion stating reasons for concluding that environmental impacts associated with the proposed release of the property is bounded by a previously issued appropriate EIS. The reviewer shall verify that the licensee has provided the following information:

- description of potential impacts on the environment, local resources, and changes to the site boundary, including site and area maps outlining impacts by types and locations

- delineation of the area slated for the PSR and the HSA basis for its non-impacted status

- description of unique site- or facility-specific issues and information not identified in prior reports submitted to the NRC

- proposed schedule of the PSR

- compliance with environmental and health protection regulations governing the presence of any other toxic or hazardous properties of materials present on the area proposed for PSR

- rationale for concluding that the environmental impacts associated with the proposed release are bounded by an appropriate previously issued EIS

If required by the presence of toxic or hazardous materials on the area proposed for PSR, the licensee should describe permits and authorizations obtained from other Federal, State, and local regulatory agencies. This aspect applies to both impacted and non-impacted areas.

2.7.4   Evaluation Criteria and Findings

Reviews performed for this section of the SRP are based on guidance and criteria listed under "Regulatory Requirements" and "Regulatory Guidance," above. The reviewer shall verify that, where applicable, the licensee's application includes the information summarized under "Information to be Submitted," above. The staff's review shall verify, to the maximum extent practicable, that the information supplied by the licensee is accurate by comparing it with prior licensee submissions, prior environmental report(s), FSAR or USAR, EISs, licensing actions, and inspection records maintained in NRC files. The staff's review shall confirm that the information supplied by the licensee is complete and demonstrates compliance with 10 CFR 50.83(b)(5) for non-impacted areas, or 10 CFR 50.83(d)(3) for impacted areas.

The reviewer shall verify that the information is provided and presented in a manner consistent with the relevant sections of NUREG-1555 or NUREG-1748. The staff shall confirm that the licensee has described the purpose of the proposed action; included a summary of pertinent Federal, State and local regulations and required permits and authorizations; identified the location and size of the area(s) proposed for PSR; identified the radiological status of the property slated for release; and identified and characterized all relevant environmental impacts. If a license amendment is required, the staff shall document its findings in either an EA (accompanied by the appropriate findings) or an EIS, as warranted. The review shall confirm whether the PSR will result in adverse effects on site environs and resources. The staff shall assess whether the proposed PSR might impact some elements of the REMP and, if so, confirm that the required changes to the program conform with Appendix I to 10 CFR Part 50 and the ODCM.

2.7.5   References

U.S. Nuclear Regulatory Commission, NUREG-1555, "Standard Review Plans for Environmental Reviews for Nuclear Power Plants, Environmental Standard Review Plan," Washington, DC, October 1999.

U.S. Nuclear Regulatory Commission, NUREG-1748, "Environmental Review Guidance for Licensing Actions Associated with NMSS Programs," Final Report, Washington, DC, August 2003.

U.S. Nuclear Regulatory Commission, Washington, DC, Regulatory Guides:
*   Regulatory Guide 4.1, "Programs for Monitoring Radioactivity in the Environs of Nuclear Power Plants," Rev.1, April 1975.
*   Regulatory Guide 4.8, "Environmental Technical Specifications for Nuclear Power Plants," December 1975.
*   Regulatory Guide 4.15, "Quality Assurance for Radiological Monitoring Programs (Normal Operations): Effluent Streams and the Environment," Rev. 1, February 1979.

U.S. Nuclear Regulatory Commission, "Radiological Assessment Branch Technical Position (BTP)," Rev. 1, Washington, DC, November 1979.

2.8   Maintenance of Records

A review of site and facility records is required under 10 CFR 50.83(a)(2), and all associated records must be maintained as indicated under 10 CFR 50.83(a). Such records are used to evaluate existing information about the area being considered for PSR, including prior and current activities, and the current radiological status of the area. This information should be used to justify the conduct of scoping or characterization surveys to address missing results, address analytical data of unknown quality or insufficient quantity, and resolve inconsistencies or information gaps in reconstructing the operational history of the site or facility from the HSA.

2.8.1   Areas of Review

In defining the area subject to radiological release criteria, the PSRregulations, under 10 CFR 50.83(a), require that all associated records must conform to the requirements of 10 CFR 50.75(g)(4). In addition to requiring that all licensees maintain records of changes in property boundaries and assess radiological conditions of current and prior licensed site areas, the requirements address records associated with PSRs from the licensed site made prior to license termination. By maintaining these records, potential dose contributions from residual radioactivity in the entire area, including any areas previously released, can be assessed in demonstrating compliance with the radiological release criteria when implementing subsequent PSRs, or when terminating the license.

2.8.2   Review Procedures

2.8.2.1 Acceptance Review

The NRC staff shall review and confirm that the application contains the information summarized under "Areas of Reviews," and described under "Information to be Submitted." The acceptance review shall confirm that the required information is complete to conduct a detailed technical analysis as the next step. The NRC staff shall confirm, via a preliminary review of the table of contents, content of each section, and any supporting attachments, that the licensee has provided the required information and determine that its level of detail is adequate. If the staff determines that the information is incomplete or insufficient to support a detailed technical analysis, the staff shall compile a list of such deficiencies and describe why the information is inadequate.

2.8.2.2 Safety Evaluation

The application submitted by the licensee should conform with regulatory requirements about the maintenance of records documenting the PSR. The staff shall review and evaluate the information at a level commensurate with the details provided in the application. The staff shall confirm that the licensee has a process to maintain (1) records of changes in property boundaries, (2) records describing the radiological conditions of current and prior licensed site areas, (3) records associated with PSRs from the licensed site made prior to license termination, and (4) records demonstrating compliance with the radiological release criteria associated with subsequent PSRs or upon license termination. The review shall confirm that the records are available, conform with requirements of the regulations, and are maintained either in specific files or as part of a distributed recordkeeping system in compliance with 10 CFR 50.83(a) and 50.75(g)(4), and that the staff has followed the review and evaluation process of 10 CFR 50.83(c) and (f) for non-impacted areas, or 10 CFR 50.83(e) and (f) for impacted areas.

2.8.3   Acceptance Criteria

2.8.3.1 Regulatory Requirements

- 10 CFR 50.75(g)(4), "Reporting and Recordkeeping for Decommissioning Planning."
- 10 CFR 50.59(d)(1) to (d)(3), "Changes, Tests, and Experiments."
- 10 CFR 50.83(a), "Release of Part of a Power Reactor Facility or Site for Unrestricted Use."

2.8.3.2 Regulatory Guidance

- NUREG-1700, "Standard Review Plan for Evaluating Nuclear Power Reactor License Termination Plans," Rev. 1.

2.8.3.3 Information To Be Submitted

The reviewer shall verify that the licensee's application package provides the following information, or that documentation confirms the availability of the information in the facility's record management system:

- definition of the licensed site, as originally licensed, which must include a site map

- any acquisition or use of property outside of the originally licensed site boundary

- licensed activities carried out on the acquired or used property

- release and disposition of any property from the licensed site, as originally licensed, or from acquired or used property added to the licensed site

- results of any HSAs performed for the disposition of such property

- scope and types of radiation surveys and survey results performed to support the release of the property

- results of 10 CFR 50.59 analyses supporting the basis and justification for the PSR

- applications made to the NRC in accordance with 10 CFR 50.83, and the methods employed to ensure that the property met the radiological criteria of 10 CFR Part 20, Subpart E, at the time the property was released

- NRC documents and correspondence granting the PSR

2.8.4 Evaluation Criteria and Findings

Reviews performed for this section of the SRP are based on guidance and criteria listed under "Regulatory Requirements" and "Regulatory Guidance," above. The staff shall verify that, where applicable, the licensee's application includes or references the information summarized under "Information To Be Submitted," above. The staff's review shall confirm, to the maximum extent practicable, that the information supplied by the licensee is complete and accurate by comparing it with prior licensing actions and inspection records maintained in NRC files. The application should be complete in documenting the required information and presented in a manner consistent with the relevant provisions of NUREG-1700. If needed, the supporting records and quality of the information may be reviewed by the staff as part of a site inspection initiated in response to the application. Licensees may maintain these records in a distributed fashion within an overall facility record management system (i.e., not necessarily contained in a specific file or folder. As stated in 10 CFR 50.75(g), if records of relevant information are kept for other purposes, references to these records and their locations may be used in documenting the existence of such records.

2.8.5 References

U.S. Nuclear Regulatory Commission, NUREG-1700, "Standard Review Plan for Evaluating Nuclear Power Reactor License Termination Plans," Rev. 1, Washington, DC, April 2003.